energy | people | buildings

MAKING SUSTAINABLE ARCHITECTURE WORK

Published by RIBA Publishing,
66 Portland Place,
London, W1B 1AD

ISBN 9781 85946 587 5

British Library Cataloguing-in-Publication Data

A catalogue record for this book is available from the British Library.

Commissioning Editor: Ginny Mills
Assistant Editor: Clare Holloway
Production: Richard Blackburn
Designed and Typeset by John Round
Printed and bound by Short Run Press Ltd, Exeter
Cover image: Judit Kimpian

www.ribapublishing.com

energy | people | buildings

MAKING SUSTAINABLE ARCHITECTURE WORK

JUDIT KIMPIAN, HATTIE HARTMAN AND SOFIE PELSMAKERS

RIBA

Contents

	Acknowledgements	VI
	About the authors	VII
	Foreword	VIII
	Introduction	1
01	The challenge	6
02	Design for people	14
03	Energy basics and benchmarks	26
04	Design for feedback	36
05	Building context and configuration	48
06	Fabric first	62
07	Integrating technical systems	76
08	Control systems and user experience	94
09	Case Studies	110
10	Contract for performance	170
	Conclusion	184
	Glossary	188
	Further reading	192
	References	194
	Index	208
	Image credits	214

OPPOSITE
Keynsham Civic Centre,
Keynsham, AHR.

Acknowledgements

This book stands on the shoulders of work by several key individuals and organisations.

For their part in the research underpinning this book we would like to especially thank:
AHR for publicly sharing lessons learned from R&D work carried out between 2006-2016; for the support of AHR's board and marketing team, as well as the project architects who undertook research on live projects with outstanding results.

Peter Oborn for his enduring support and guidance; for paving the way for embedding research in practice; for his deep interest in this work and his ongoing persuasion of industry decision makers to recognise the value of building performance feedback.

Harry Bruhns, Bill Bordass, John Field, Rob Cohen and Hywel Davies for all the mentoring; for their role in identifying the performance gap; for sharing their insights on the importance of tracking data over a building's lifespan and much more.

Professor Dejan Mumovic for his passion for engaging industry with academic research and for championing the studies that inform this book.

Dr Esfandiar Burman, for sharing his ground-breaking building performance research and evaluations of AHR buildings; for his insights into the complex causes of the 'performance gap' and for his in-depth review of chapters 7 and 8.

UCL/AHR PhD/EngD researchers for their excellent work on AHR/UCL research projects: Dr Craig Robertson, Dr Craig Paterson, and Jamie Bull.

AHR R&D team: Dan Rigamonti for pulling out all the stops to make research relevant to practitioners and apply the results; Sophie Chisholm for her excellent work on AHR's BPE projects and CarbonBuzz; Josh Mason for the inspiration to substantiate the term 'sustainable' with evidence.

Gary Clark and the RIBA Sustainable Futures Group for spearheading the RIBA Climate Challenge that has transformed industry discourse in a few short months.

Colleagues at the Architects' Council of Europe for enthusiastically embracing the policy implications of building performance research.

UKRI for funding Building Performance Research to highlight the critical impact of the performance gap.

In addition, we would like to thank:
The practices and clients who went above and beyond to supply the supporting evidence and images for the case studies by responding to our endless questions: Dr Craig Robertson, AHMM; Christian Dimbleby and Gareth Selby, Architype; Sarah Lewis, Hansom Barron Smith; Steve Tompkins, Will Mesher and Diana Dina, Haworth Tompkins; Adam Spall, AHR, Dr Esfandiar Burman, UCL, Alasdair Donn, Willmott Dixon, Tamsin Tweddell Max Fordham, and Jeff Tatum, Bath & NE Somerset Council; Alasdair Donn, Willmott Dixon, Jakob Strømann-Andersen + Imke Wies van Mil, Henning Larsen; Bailey Thompson, ZGF Architects, Robert Longthorne, Everyman Theatre, and Kelly Vaughn and Joey Cathcart, Rocky Mountain Institute.

The generous (many anonymous!) reviewers of the draft manuscript who offered insightful comments that enabled us to extensively revise the book during lockdown. In addition to names mentioned above, special thanks to peer reviewers Fionn Stevenson, Mina Hasman. and Robin Nicholson whom we also thank for his thoughtful Foreword. And to Anastasia Mylona and Julie Godefroy at CIBSE for their prompt clarifications and to Azlizawati Ibrahim for support documenting references.

Heidi Sukanen for her clarity and creativity translating technical concepts into concise and accessible illustrations. Tampere University for co-sponsoring the graphic design of the diagrams. RIBA Publishing for supporting us throughout the evolution of this book.

This book had a long gestation and we are immensely grateful to our partners for keeping us sane. Special thanks to Federico Calboli for a massive editing job during the final push and for help with references; to Alasdair McPhail for his insights into the important role of UX in product design and its absence in architecture, as well as help with 3D cutaway sections.

JK, HH and SP

About the authors

Architect and environmental policy expert **Judit Kimpian** chairs the Architects' Council of Europe's (ACE) Sustainable Architecture Group, where she leads a campaign to reform EU legislation to deliver a net zero built environment by campaigning for revision of EU policy, directives and standards.

During 20 years in practice, including five years as Director of Sustainable Architecture and Research at AHR, Judit was responsible for the realisation of award-winning international design projects. She also led collaborative research, bringing together industry and academic partners to tackle the gap between the expected and achieved performance of buildings, and undertook numerous performance studies of completed buildings in use.

The insights she gained into the impact of feedback on design outcomes led her to engage stakeholders across the construction sector to benchmark and share performance data and lessons learned in the public domain. Her work has informed the development of CIBSE and RIBA online tools and technical guidance. Judit teaches energy efficient architecture and planning at UCL and supports organisations with embedding feedback in policy, design and construction. She regularly contributes to professional publications and is a frequent speaker at national and international conferences.

Journalist and author **Hattie Hartman** is sustainability editor at the *Architects' Journal*, London. Raised in America and trained as an architect and urban planner at Massachusetts Institute of Technology, she has lived in London since 1991. After 15 years in practice, serendipity and the opportunity to interview Oscar Niemeyer led her into journalism.

She joined the AJ as technical editor in 2006, creating the post of sustainability editor two years later. In 2018, Hattie was named an RIBA Honorary Fellow for helping shape the profession's agenda. She lectures frequently on design for climate emergency and regularly chairs industry panels and juries. In 2020, she launched *Climate Champions*, a podcast from the Architects' Journal in association with ACAN, the Architects Climate Action Network.

Hattie is the author of London 2012 *Sustainable Design* (Wiley, 2012) and editor of *Invisible Architecture: Atelier Ten* (Laurence King, 2015). She guest-edited *AD 241 Brazil: Restructuring the Urban* (Wiley, 2016).

Sofie Pelsmakers is an architect and assistant professor in Sustainable Housing Design at Tampere University, Finland. As an environmental architect, educator and researcher on affordable, low energy housing design, she is interested in how buildings and places work in reality over time. This includes how building use and users change and adapt buildings, and how architecture affects building use and users. She brings experience from the UK and Denmark, and now Finland.

Sofie authored the *Environmental Design Pocketbook* (RIBA, 2nd ed. 2015), a comprehensive publication which distils environmental science, legislation and guidance into one source, accessible for both practitioners and students. This publication received commendation by the RIBA President's Medals for 'Outstanding Practice-located Research' (2012) and was 'Highly Commended' for the UKGBC/PRP 'Rising Star award' in 2013. Among other publications, Sofie guest edited the RIBA Design Studio '*Everything Needs to Change: Architecture and the Climate Emergency*' (RIBA, 2021) and is co-authoring a forthcoming student guide to designing for the climate emergency. She was selected as an RIBA national Role Model (2015) and was one of two Bartlett PhD Alumni Role Models (2017).

Foreword

Why has it taken so long for architects, all of us, to recognise the importance of feedback? Since first reading the outline for this book, I was impatient for it to be published and become the must-read book our industry so desperately needs. The three authors are all architects: one a practitioner with research interests, one an academic, and one a journalist covering environmental matters. As they say, this book presents actual feedback from buildings in use, and explains why 'feedback is a transformational tool that can give designers the confidence to deliver low-carbon buildings'.

Developing the feedback habit has been a long, slow journey over the 50 years since I was a postgraduate at the Bartlett, a school of architecture that taught environmental design to architecture students. But that was only for the undergraduates, so it didn't really influence those of us who had come from Cambridge or the USA. Nevertheless, as we began to challenge established norms, we understood the importance of research. Reyner Banham was writing his ground-breaking *The Architecture of the Well-Tempered Environment* (1969), and he introduced us to cultural changes in California. We devoured alternative publications such as Stewart Brand's *Whole Earth Catalog* (1968–72).

Under the inspired leadership of RMJM partner Andrew Derbyshire, the RIBA first published the Plan of Work in 1964, and it included Part L for feedback, the focus of which was occupant feedback rather than any quantitative metrics. However, that was quickly dropped, and only reinstated in 2013. Energy was so cheap that its conservation was of little interest, and the impact of greenhouse gases was not even discussed. It took the 1973 oil crisis to start to focus designers on reducing energy, and 1976 saw the first regulations. We had to wait until 1987 for the Brundtland Declaration and 1992 for the Rio Earth Summit to make greenhouse gas emissions a global responsibility.[1] *Energy, People, Buildings* makes the point that energy use in buildings should not only be measured to combat climate change, but because it is an overarching metric that helps diagnose issues with build quality and operation.

It should be a matter of routine best practice for architects to want to learn from what they have designed so they can do it better next time, but existing professional indemnity schemes discourage architects from finding out what could be improved, which means that mistakes keep getting repeated. One welcome, if slightly shocking, aspect of this book is how responsible architects are open about sharing lessons learned when things go wrong. Even though the case for this has been made repeatedly, there is no obligation and seldom any funding for feedback, while design fees continue to fall. This book explains how to overcome these challenges.

RIBA President Frank Duffy (1993–95) always argued that, as a profession, architects were becoming more ignorant by failing to undertake feedback as a matter of course. However, the higher education sector is one exception, and the HE Design Quality Forum publishes excellent guidance on feedback. Vice-chancellors take the People and Planet University League annual surveys very seriously.

As a practice we always tried to get feedback, even if only anecdotal. Occasionally, as at the Lambeth Community Care Centre (1985), we helped with formal feedback on both occupant satisfaction and quantitative performance metrics. But it wasn't until the University of Cambridge agreed that our Centre for Mathematical Sciences could be the subject of a Post Occupancy Review of Building Engineering (PROBE) study in 2001 that we fully appreciated the importance of the learning that emerges from such a rigorous process. Both the university and we were nervous about the outcome, especially since the second phase was on site, with all the disruption that can cause. We needn't have worried, though, as the building works well, scored in the top percentile and is gratifyingly popular with its users.

The PROBE studies (1995–2002) undertaken by Bill Bordass and Adrian Leaman with other colleagues and published in the *CIBSE Journal* pioneered a simple and straightforward approach to feedback. As a building scientist and a social geographer respectively, Bordass and Leaman were concerned about the performance gap and its implication for professional responsibility, a line of enquiry which became central to the Edge think tank. Following two Edge debates on the subject, Bordass and Leaman edited a special issue of *Building Research & Information*[2] in 2012 that itemised ten tenets of a new professionalism. The sixth is 'Evaluate and reflect upon the performance in use of your work. Feed back the findings.' In a similar vein, *Architects Declare* (2019) advocates 'incorporating post occupancy evaluation as part of our basic scope of work'.

Energy, People, Buildings builds on this work and adds to a growing number of excellent papers and books on feedback that have been published recently. The London Energy Transformation Initiative (LETI) *Climate Emergency Design* Guide,[3] written by over 100 volunteers and launched in January 2020, includes an informative chapter on data disclosure, and illustrates the growing desire among younger professionals to bring about change. *Energy, People, Buildings* goes further in explaining both the physics and the human dimensions, the metrics and the procurement. It operationalises feedback with a five-stage cycle of Plan, Design, Procure, Use and Learn, and makes it easy to see who has to do what, and when.

Despite the growing practice of feedback, the authors argue that there are still relatively few genuinely low-energy buildings with data available for publication. Seven case studies cover a range of building types, including one in Denmark and one in the USA, with data, drawings and photographs. They are primarily newbuild, with one rebuild and one retrofit/extension.

This manual is beautifully written, with clear diagrams and comprehensive references after each chapter. Just when you are feeling energised to do more and better, the authors conclude with a chapter on procurement that spells out how to make feedback happen in practice. As our industry inches from regulatory compliance towards performance verification in the coming years, we will all need to adopt this approach. The UK government is slowly working its way towards instituting Dame Judith Hackitt's 'golden thread' in the wake of the Grenfell tragedy. A similar approach is needed to deliver a net zero built environment, and *Energy, People, Buildings* explains how to make this a reality.

COVID-19 has temporarily disrupted much of the world's complacent consumption, but personal, political and economic pressures are driving traffic and air quality back to pre-crisis levels. Two years on from her initial school strike, Greta Thunberg wrote in the *Guardian* that 'the willingness to act and the level of awareness needed are still nowhere in sight'.[4]

Energy, People, Buildings provides both the inspiration and a guide for the industry to act. We must better understand our buildings and change our ways. If every building design practice devoted one hour a week to reading one chapter and then discussed their work through that lens, preferably with their favourite engineer and client, then in just ten weeks we would be a much wiser industry.

ROBIN NICHOLSON CBE RIBA Hon FIStructE Hon FCIBSE
Cullinan Studio partner 1980–2020 and convenor of the Edge.
See www.edgedebate.com

Introduction

This is not just another theoretical book about energy performance in buildings. What makes this book different is that it focuses on evidence from actual feedback from buildings in use. It explains what it takes to get buildings to perform, and the reasons why they often fail.

It is a handbook for designers to understand the factors that impact people's comfort and satisfaction and the energy use of buildings, and to track them across the life of a project to ensure that buildings perform as intended.

Energy, People, Buildings draws extensively on the findings of research, UK government-funded building performance evaluations and on original research into seven case-study buildings that achieved exemplary performance through the use of building performance feedback.

The book illustrates that feedback is a transformational tool that can give designers the confidence to deliver low-carbon buildings in use. It also argues that it is essential in the face of the current climate crisis.

FIGURE 0.1: OPPOSITE
Frederiksbjerg School, Aarhus, Henning Larsen Architects and GPP Arkitekter.

How to use this book

This book is structured in ten chapters.

> **'MOST IMPORTANT IS TO LEARN FROM THINGS YOU DID WELL AND FROM YOUR MISTAKES AND WORK OUT HOW TO DO THEM BETTER NEXT TIME.'**
>
> ANNALIE RICHES, MIKHAIL RICHES[1]

Two introductory chapters set the scene. The first chapter describes the challenge of realising zero carbon buildings in response to the climate emergency. Chapter 2 explains the important role of people in the energy consumption of buildings, and why so often things go wrong because the human factor is poorly understood at design stage. Chapter 3 provides an overview of energy basics, i.e. what a designer needs to know to be energy 'literate'. Chapter 4 sets out feedback processes and ways to overcome barriers to feedback.

The book's middle chapters (5 to 8) review the factors that determine building performance, such as passive design (chapter 5), building fabric specification (chapter 6), technical systems (chapter 7) and automation and controls (chapter 8).

Chapter 9 showcases seven case studies that draw out the issues explained in the previous chapters and illustrate how building performance works in practice. The case studies can be read alongside the earlier chapters.

Finally, chapter 10 outlines innovative approaches to incorporating feedback into project procurement, and focuses on the use of a Building Performance Register (BPR) to track a building's performance from design to operation. It describes what it takes to align supply-chain incentives for high building performance so that client, design team and contractor work as a collaborative partnership to deliver buildings that work as intended.

As this is a 'how to' handbook, actions for designers are summarised at the end of chapters 5 to 9 and are grouped across five key stages: Plan, Design, Procure, Use and Learn. The focus for each stage is as per Figure 02.

FIGURE 0.2
The five key stages of the feedback process: plan, design, procure, use and learn, aligned with the RIBA Plan of Work.[2]

ACTIONS	PLAN	DESIGN
	• Agree explicit and realistic performance goals at briefing stage.	• Calculate and optimise design performance based on accurate assumptions and likely operating scenarios: i.e. quantifying, qualifying and refining initial performance goals.

A call to action

Energy, People, Buildings demonstrates how continuous feedback loops from the briefing, design, construction and operation stages of buildings can give greater accountability in a project's life cycle, reduce performance risks and inspire a new type of architecture. Choosing the right monitoring strategies and agreeing early in a project cycle how these will be reported can be transformative for user comfort and satisfaction, and the operational energy consumption of buildings.

It is our hope that this book will help to demystify the complex subject of occupant satisfaction, comfort and energy use in buildings. Understanding all these aspects enables designers to balance human factors and environmental impacts across the life-cycle stages of buildings with confidence.

Given that reducing the impact of all resource consumption is crucial to achieve low-impact development, a building's embodied carbon has rightly become a recent focus. Embodied carbon will become ever more important because it increases proportionally as energy-efficiency measures drive operational energy use down. To reach zero carbon, designers must ultimately address whole-life carbon – both operational and embodied.

Yet we are far from solving the issue of operational energy use, and an in-depth understanding of building performance is essential for architects to contribute to the low-carbon transition – this is the focus of this book.

PROCURE	USE	LEARN
• **Embed performance targets in construction and commissioning contracts.**	• **Validate achieved performance in use, support early building diagnostics comparing initial goals with measured performance in use to rectify problems early.**	• **Share lessons for continuous improvement on future projects and in the wider sector.**

A note on terminology in the book

Numerous terms are used interchangeably when referring to building performance, but each mean something slightly different. The most common are post-occupancy evaluation (POE), building performance evaluation (BPE) and the term 'feedback'. These are explained briefly below.

POE commonly refers to studies that are undertaken post-building completion when a building is in use, and may include both quantitative and qualitative information such as energy monitoring, studies of indoor environmental quality, comfort, building design, surveys or interviews with building occupants and/or facilities managers, and building walk-throughs.

BPE is more comprehensive and includes POE, and is undertaken at any point in a project's cycle, but preferably throughout design, construction and building operation. It involves observations and qualitative feedback from occupants relating to comfort, functionality and wellbeing, alongside energy consumption and indoor environmental quality data. The evaluation of a project's resource consumption, including embodied carbon, material impacts, climate-change resilience and life-cycle costs are also increasingly considered.

Feedback is a more general term that covers both POE and BPE, and refers to the process of evaluating and improving performance throughout all project stages. Like BPE and POE, it implies corrective action that should be taken post-completion.

Performance gap is another important term in this book. It is often used to refer to the difference between predicted (or calculated) building performance at design stage based on modelling, and actual measured performance based on meter readings once the building is in use. In this book, the performance gap is used more broadly to also include all aspects of a building's performance that relate to people, such as comfort and occupant satisfaction.

In this book, the term 'feedback' is used as a general term to include all types of feedback processes, including POE and BPE. The use of the term 'building performance' or 'performance' refers to a building's technical performance reflected by energy use, but also the human performance such as occupant comfort and user satisfaction. Furthermore, the terms 'zero energy' and 'zero carbon' are used throughout to encapsulate terms such as 'nearly zero energy building' (nZEB), 'net zero' etc. – see page 29.

The first instance of any underlined words are further explained in the glossary at the back of the book.

FEEDBACK IS A MORE GENERAL TERM THAT COVERS BOTH POE AND BPE, AND REFERS TO THE PROCESS OF EVALUATING AND IMPROVING PERFORMANCE THROUGHOUT ALL PROJECT STAGES. LIKE BPE AND POE, IT IMPLIES CORRECTIVE ACTION THAT SHOULD BE TAKEN POST-COMPLETION

A note about case studies

This book draws extensively on examples and lessons learned from UK government-funded building performance evaluations (BPE), led by AHR R&D together with UCL IEDE.

Chapter 9 presents seven detailed case studies specifically collated for this book, which illustrate how design with performance in mind makes sustainable architecture work, through real life examples.

The case studies are signposted across the chapters. These are:

1 Everyman Theatre, Liverpool by Haworth Tompkins Architects
2 Morelands Rooftop, London by AHMM
3 Carrowbreck Meadow, Hellesdon, Norfolk by Hamson Barron Smith
4 Rocky Mountain Institute, Basalt, Colorado by ZGF with Graybeal Architects
5 Frederiksbjerg School by Henning Larsen and GPP Arkitekter, Aaarhus, Denmark
6 Keynsham Civic Centre, Keynsham, Somerset by AHR
7 Enterprise Centre, University of East Anglia, Norwich, Norfolk by Architype"

The BPE buildings, all designed by AHR, are referenced in the text as follows:

- School 1 – Loxford School of Science and Technology;
- School 2 – Brine Leas School;
- Academy 1 – Pennywell Academy;
- Academy 2 – Petchey Academy;
- Academy 3 – Stockport Academy;
- Innovation Centre 1 – Pool Innovation Centre;
- Innovation Centre 2 – Tremough Innovation Centre.

Most of the examples in this book are non-domestic buildings. However the majority of key lessons drawn from non-domestic research and case studies also apply to housing and other building types.

01

THE CHALLENGE

FIGURE 1.1: OPPOSITE
Everyman Theatre, Liverpool, Haworth Tompkins Architects.

Introduction

This chapter sets out the roots of the performance gap and underlines the central role of people in buildings in order to address it. This sounds obvious, but the human factor is often unintentionally overlooked.

Validation of performance in use offers a clear route to simultaneously transform the way buildings are designed and built, and lighten their footprint on the planet. The way people live in and use their buildings is the starting point for this change.

In the face of the dual challenges of a global pandemic and the climate emergency, extensive policy discussion is focused on resetting economies towards a green recovery. Transparent sharing of lessons learned from studying buildings in use can help to accelerate this industry change. In light of feedback gathered from buildings in use, a re-examination of the assumptions that drive sustainable building design can help to ensure that the long-term needs of people and the environment are at the heart of this transition.

This chapter also hints at the need for a procurement revolution to tackle this challenge head-on – a theme that surfaces repeatedly throughout the book.

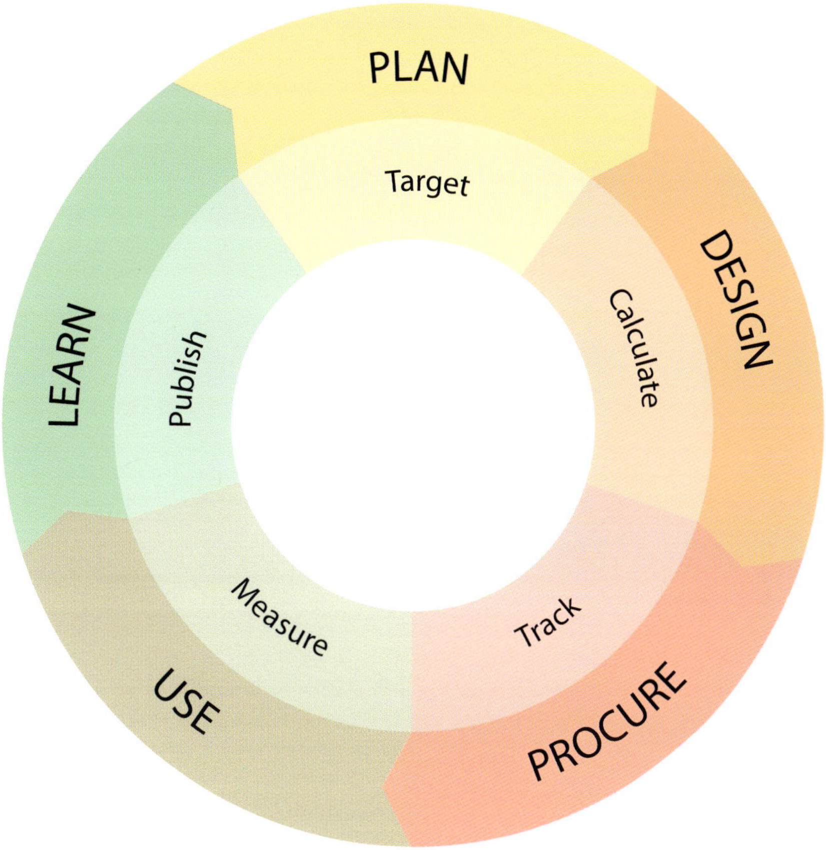

FIGURE 1.2
Five key stages of feedback.

The performance gap

Concern about energy use in buildings emerged initially during the 1970s oil crisis, culminating in 2019 when architects put forward a global declaration[1] outlining the actions required to design buildings that use less energy and fewer resources while offering better-quality environments.

However, neither this past awareness nor ever tighter building regulations have translated into significant reductions of CO_2 emissions in the building industry (see figure 1.3). Improvements in fabric performance and system efficiency have been offset by increased floor area, and rising demand for cooling and running technical equipment in buildings.[2] To date, reductions in CO_2 emissions have been mainly accomplished through the gradual decarbonisation of electrical power at country level.

Buildings with ambitious sustainability credentials multiplied across the globe with the increased momentum of the green building movement that led to the creation of certification systems such as LEED and BREEAM in the 1990s. Initially, these rating systems mainly focused on the reduction of energy use, and green buildings were designed with this objective in mind.

FIGURE 1.3
In 2019, the UK Committee for Climate Change reported that despite ever tighter building regulations, reductions of CO_2 emissions in buildings have been marginal.[3]

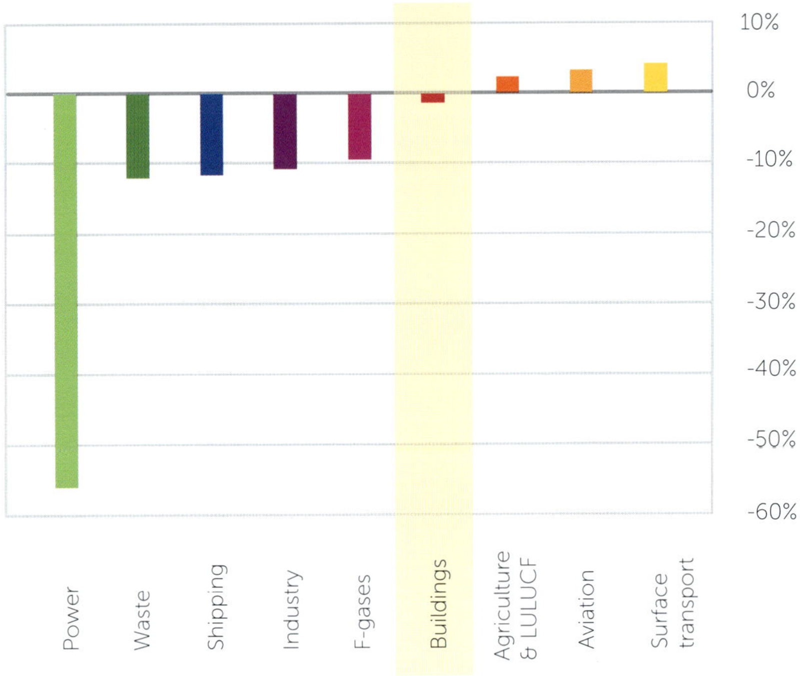

NEW NON-DOMESTIC BUILDINGS AND MAJOR REFURBISHMENTS IN THE UK FREQUENTLY CONSUME 1.5 TO 2 TIMES MORE ENERGY THAN ANTICIPATED, WHILE FOR NEW DOMESTIC BUILDINGS ACTUAL CO_2 EMISSIONS CAN BE UP TO 2.6 TIMES AS MUCH.

Yet in 2006, the UK's implementation of Display Energy Certification (DEC) revealed an alarming gap between energy use predicted at the design stage and actual energy use once buildings were occupied (see figure 1.4). This discrepancy is referred to as the 'performance gap'. New non-domestic buildings and major refurbishments in the UK frequently consume 1.5 to 2 times more energy than anticipated,[4] while for new domestic buildings actual CO_2 emissions can be up to 2.6 times as much.[5]

Many new buildings also underperformed in terms of indoor air quality and thermal comfort (particularly summer overheating), often due to complex technical systems that were not intuitive to use and did not operate as intended.[6] This widespread phenomenon has spurred the industry to identify the scope and causes of the performance gap.

Analysis of performance gap data found that sustainability measures set out at the design stage frequently did not make it through to the finished building as intended.[7] DEC data analysis uncovered problems across all project stages, including unrealistic design expectations, poor construction quality, a highly fragmented supply chain and unexpected operating conditions. Such underperforming buildings create a long-term investment risk and hinder the ability to meet climate targets.

SIGNIFICANT IMPROVEMENTS WERE ACHIEVED IN PROJECTS WHERE PERFORMANCE VALIDATION WAS A CONTRACTUAL REQUIREMENT.

While the majority of projects performed below expectations, significant improvements were achieved in projects where performance validation was a contractual requirement (see case studies 4, 6 and 7). Globally, initiatives that introduced some contractual obligations of expected performance, such as the UK's DECs, Germany's Passivhaus, America's Energy Star and Australia's NABERS, gained traction to address the performance gap, and this facilitated the adoption of construction practices, which produce predictable results that can be validated.

Regulatory framework

The regulatory framework designed to improve building energy use was found to have numerous shortcomings and unintended consequences.[8] Instead of targeting actual measured performance, it only requires the calculation of a building's likely energy use under idealised conditions, perfectly constructed and operated, and largely without occupants or the equipment that they bring.

As a result, investment in energy efficiency often fails to deliver the expected improvements, and CO_2 emission reductions are only achieved on paper. The performance gap is shown by figure 1.4. This leaves building owners and managers, not to mention end users, frustrated due to excessive maintenance and operating costs and amelioration works, undermining confidence in the low-carbon construction sector. Without any reliable indicators of energy consumption in use, there is no evidence base to link a building's performance to a building's value. As a result, neither effective financial incentives, nor statutory instruments, have had a chance to evolve.

Meanwhile, reliance on complex building technology has increased exponentially on the assumption that automated or smart technical systems will help to reduce consumption. Despite its potential, technology disproportionately contributes to higher-than-expected

FIGURE 1.4

Diagram showing the difference between calculated energy performance for regulatory purposes (EPC) and actual measured performance (DEC) in Innovate UK's BPE programme.[9]

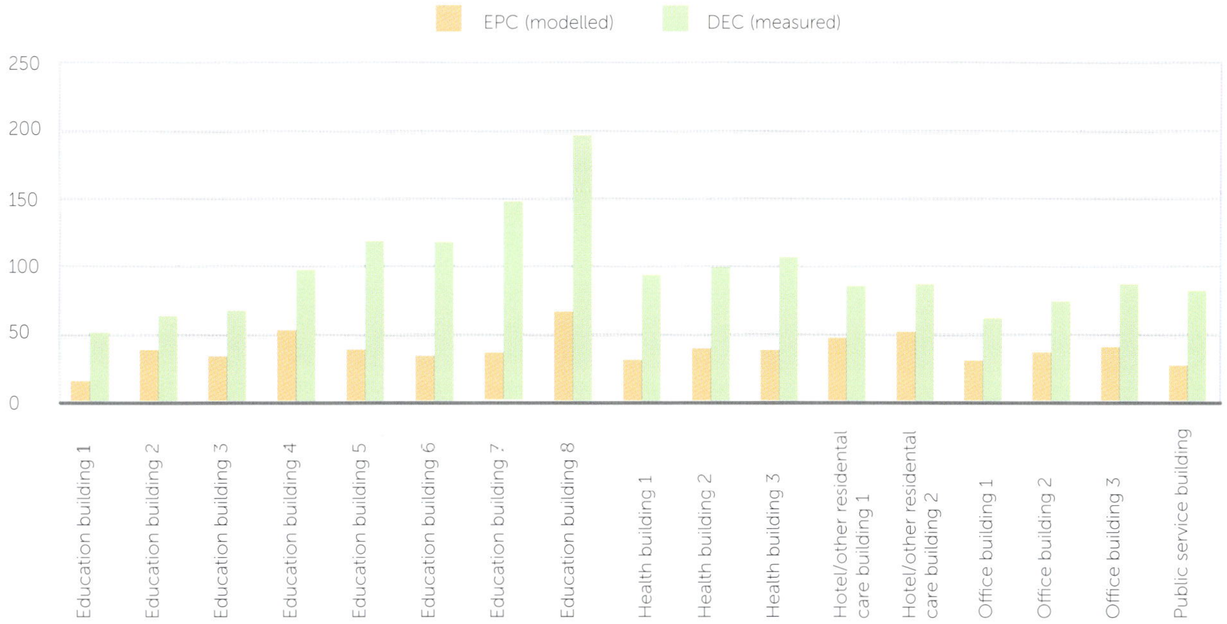

energy consumption and maintenance costs in buildings.[10] Technical complexity increases the risk of failure in the absence of any regulatory oversight of control systems or the obligation to ascertain if these are delivering the energy performance anticipated.

At the same time, energy efficiency measures without careful consideration of indoor environmental conditions can contribute to overheating, poor air quality and insufficient daylight.[11] Moreover, efforts to reduce operational energy use frequently increase the embodied carbon impact of buildings,[12] and in some cases compromise long-term safety (e.g. fire risk) and adaptability. While a holistic approach to all these issues is critical for good building design, a thorough understanding of performance is an essential starting point.

To that effect, momentum is building in some EU countries with the introduction of reporting requirements of whole-life carbon emissions, such as in France, Finland and the Netherlands. At the time of writing, the European Commission is considering mandating operational energy reporting as part of its legislative drive to improve the energy performance of buildings. And professional bodies such as the RIBA and the American Institute of Architects (AIA) have issued climate challenges to their members, setting out a trajectory towards zero carbon buildings that is based on actual performance and evidenced by measured data.[13]

Procurement

Realising green buildings requires a coordinated inter-disciplinary effort, and a willingness to take risks and learn from mistakes. This is difficult under mainstream adversarial procurement routes, which incentivise teams to work in isolation and share only the required contractual information. If clear lines of responsibility for realising performance are missing, compliance with legislation becomes the maximum rather than the minimum to be aimed for. By law, project teams only have to take responsibility for 'notional' building performance, not achieved performance in use, which further detaches designers from their finished product and end users.

Under design and build (D&B) contracts, novated consultants are unable to report directly to the client should the contractor make any changes to the design that would affect design quality and performance. Designers have little power to prevent cost engineering of their design, which means that a building may no longer function as intended (e.g. the omission of critical design elements, or substitution with low-cost versions).

Employment of named specialist subcontractors is often necessary to successfully install innovative technologies. However, this is not possible under a D&B contract, where each item must tender to several suppliers. This is where the procurement of projects requiring validation of in-use performance can be transformational.

1.01: EXAMPLE

IMPACT OF VALUE ENGINEERING ON PERFORMANCE

Automated natural ventilation systems often suffer from incompatible actuators that jam because the proprietary actuator was 'value engineered' for a cheaper generic version.

The human factor

How people use and experience a building, and how convenient it is to navigate and access spaces, and operate the building's 'interfaces', has a major impact on occupant satisfaction and comfort. Yet, the growing disconnect between designers and end users has been an important factor in the construction sector overlooking the importance of user-centric design. Unrealistic assumptions about workmanship, occupancy patterns and maintenance requirements have lingered due to the lack of robust feedback mechanisms from completed buildings in use. Clearly, feedback about actual operating conditions is not an impediment but a rich resource to inform design and continuous improvement.

UX (user experience) design played a important part in the realisation of all of the case studies featured in this book. This was achieved through strong leadership, enabling project teams to go far beyond legislative requirements and to overcome procurement challenges. Committed leadership is one of the most important success factors for well-performing buildings, fostering professionalism in project teams and maintaining clear lines of responsibility.

The role of the architect in feedback

The technical aspect of building performance has led to the perception that feedback is a technical pursuit to be undertaken by engineers, independent from an appraisal of architectural quality and from how people actually inhabit and use buildings. This divide is exacerbated by legislative frameworks that often focus only on the quantifiable aspects of building performance.

In fact, feedback processes offer architects an opportunity to take part in this debate. Well-conceived building performance studies with appropriate feedback loops can reconnect designers with a building's users and foster a more humanistic approach to sustainable design. Empathetic design of low-energy environments with people at their centre will result in better buildings that owners will look after and occupants will cherish. The case studies in chapter 9 point the way forward by demonstrating what is possible when feedback is used effectively.

Encouragingly, a 2018 survey by the Architects' Council of Europe found that 19% of UK architectural firms and 13% in the EU claimed to have engaged in some form of post-occupancy evaluation at some point. However, clearly there is still a long way to go.[14]

CLEARLY, FEEDBACK ABOUT ACTUAL OPERATING CONDITIONS IS NOT AN IMPEDIMENT BUT A RICH RESOURCE TO INFORM DESIGN AND CONTINUOUS IMPROVEMENT.

1.02: EXAMPLE

IMPACT OF CONTRACT ON PERFORMANCE

A tried-and-tested automated window system fitted in a traditionally procured building could not be supplied on a Design and Build project because of the exclusion of a named supplier. Subsequent problems with integrating controls and actuators from multiple suppliers adversely impacted the building's performance.[15]

1.03: EXAMPLE

OCCUPANCY PATTERNS AND USAGE PROFILES AFFECT ENERGY USE

Technology has changed how we work and live in buildings. Even before the unprecedented impact of the coronavirus pandemic, mobile technologies had prompted ever more transient occupancy patterns, with hot-desking and remote working becoming more common.
The transformation of the workplace from formal office working hours to round-the-clock access, and the hosting of activities outside of a standard usage profile are increasingly common. For example, higher education campuses are moving towards 24-hour building access. In housing, occupancy patterns are also changing, as home working and sharing of accommodation both become more common. Intermittent and out-of-hours occupancy that is not sufficiently planned for is one of the most common causes of high energy consumption in buildings.[16]

OUSE

02

DESIGN FOR PEOPLE

FIGURE 2.1: OPPOSITE
JTP offices, Pennington Warehouse, London, JTP

Introduction

This chapter looks in detail at the human aspect of building performance: people's comfort and satisfaction. In the context of a changing climate, comfort parameters must be reassessed. Today's challenge is to create comfortable environments while minimising impact on the planet.

The strength of feedback is that it shows what actually works in practice, helping to streamline both the design and construction processes. Information about occupant comfort can be gleaned both from quantitative metrics gathered from sensors, and from qualitative parameters collected through surveys, interviews and meetings.

In order to undertake feedback effectively, design targets and indicators must be clear and comparable to information collected at the in-use stage.

Perhaps unsurprisingly, occupant feedback can be gathered much more efficiently, and hence with lower fees, if planned for from the outset of a project.

Human wellbeing

Most energy in buildings is used to provide occupants with thermal, respiratory and visual comfort; however, other aspects of indoor environmental quality (IEQ) also affect occupant wellbeing. Many IEQ indicators are quantifiable, and indoor air quality (IAQ) and thermal comfort are the most frequently monitored of these.

The most common parameters that affect IEQ relate to human senses and perceptions (table 2.1), and are discussed in more detail in this chapter.

Other, less quantifiable, aspects of quality include tactility, functionality and usability, or spatial and design qualities, accessibility and safety. How spaces are used, their density and pattern of occupancy, will impact on human comfort, productivity and energy use.[1] And some IEQ parameters affect one another – for example air movement affects air quality as well as thermal comfort.

The kind of parameters that are set as targets and tracked from design to operation can make a big difference to both indoor environmental quality and the energy consumed to achieve it. Table 2.1 Common parameters

TABLE 2.1: COMMON PARAMETERS AND INDICATORS AFFECTING IEQ[2]

IEQ aspects	Qualitative parameters and quantitative indicators or metrics
Thermal comfort	• Temperature (°C) • Humidity (%) • Overheating hours • Radiant heat • Air flow (m/s) • Metabolic rate • Clothing customs
Respiratory comfort/IAQ	• Outdoor pollutants: NO_2, PM2.5, PM10 • Indoor pollutants: CO_2 concentrations, volatile organic compounds (VOCs), persistent organic pollutants, formaldehyde. See note 2.02 • Air changes per hour (ach^{-1}) or per person/second
Visual comfort	• Lighting levels (lux) • Colour temperature (Kelvin) • Daylight factor (%) • Contrast ratio (glare) (%) • Sky and planting in the field of view (%) • Variability of lighting levels • Background vs activity lighting
Acoustic comfort	• Noise levels (dB) • Reverberation time

FIGURE 2.2
JTP offices at Pennington Street Warehouse, London. The architects used 'before and after' staff surveys to understand how to support more collaborative working when the practice moved to newly refurbished premises. See also note 2.01.

and indicators affecting IEQ. Quantifiable indicators and recommended and safe levels can be obtained from the World Health Organization (WHO) or country-specific codes and guidance.

When occupant and neighbourhood needs are thoughtfully considered during design, they can inspire architectural expression, strengthening local identity and contributing to a community's sense of being valued – see note 2.01.

In the past decade, the twin topics of human comfort and indoor environmental quality and their relationship to productivity gains were explored, particularly in the office and education sectors. Both wellbeing and productivity are highly subjective terms, yet ample evidence has shown that poor IEQ causes a drop in cognitive performance and productivity and can lead to health issues, while good comfort conditions improve concentration, clarity of thought and strategic thinking in both domestic and non-domestic settings.[3] COVID-19 has further magnified the need to rethink the health implications of contemporary approaches to indoor health.

Occupant satisfaction

Occupant satisfaction is the degree to which occupants prefer or dislike different aspects of IEQ. It can only be 'measured' through surveys and interviews.

Non-building-related factors such as the quality of organisational management, the novelty of new premises, or even location can affect occupants' perception of comfort and wellbeing. Methodologies for correcting for these variables when carrying out occupant satisfaction surveys are being developed.[4] To realise low-energy buildings, a better understanding of the role of architecture in delivering occupant perception of comfort, health and productivity is needed.

Feedback studies confirm that occupant perception of comfort does not necessarily match measured IEQ. For example, in buildings where occupant needs in terms of design and functionality are met, the comfort range for both temperature and air quality can be greater than in buildings where people are not satisfied with the design. This is referred to as the 'forgiveness factor' – see note 2.03.[5] Equally, when an aspect of comfort is significantly out of range, it can negatively impact the perception of other parameters and affect productivity.

The impact of design quality on other performance parameters is rarely assessed, but there is a growing interest in understanding and communicating the relationship between design quality and quantifiable social outcomes.[6] 'Before and after' surveys are a compelling way to communicate the value of design, and can provide evidence about a building's design according to a wide range of factors, including:[7]

- productivity gains
- attendance rates
- staff sick days
- medication and forced fixation rates (psychiatric hospitals)
- exam completion rates (schools)
- extended life span (elderly people's residences)
- crime rate
- asthma rate
- property value.

The auditing of actual environmental performance has found that the interaction between people and buildings is a major driver in occupant satisfaction.[8] Standard IEQ or workplace surveys usually include qualitative statements from users that can be helpful in interpreting statistical feedback about people's perception of their environment. Moreover, surveying occupant satisfaction can in itself bring about higher satisfaction rates, especially if those surveyed perceive their feedback to be acted on.

2.01: EXAMPLE

THE VALUE OF FEEDBACK

Placemaking practice JTP relocated to a refurbished warehouse in early 2019 to support a more flexible and collaborative project-based approach to working (see figure 2.2.). Extended staff surveys were conducted before the move to understand everyone's needs. After the move, in-use energy and IAQ monitoring were undertaken alongside an 86-question staff survey. A year later, 95% of employees rated the new studio as an environment more conducive to project-based working, while 85% felt empowered to move around the studio and work in different zones. Issues that were highlighted for ongoing improvement include draughts in certain areas and considerable temperature variation between floors. Remedial actions include identifying and rectifying the source of draughts as well as resetting the Building Management System (BMS) for monitoring of temperatures and relative humidity to optimise control settings.

2.02: EXAMPLE

MONITORING IAQ

CO_2 levels can be an important indicator of whether ventilation systems are working as intended (see case study 6, and chapter 8). Shortcomings in IAQ are not necessarily noticeable to occupants, and energy consumption on its own is often not enough to reveal them. When IAQ parameters are automatically tracked in operation, issues are revealed early on.

2.03: DEFINITION

FORGIVENESS FACTOR OR FLEXIBILITY SCORE

The 'forgiveness factor'[9] originates from feedback studies[10] where people were found to tolerate more uncomfortable conditions in naturally ventilated buildings than in fully sealed buildings. 'Flexibility score' may be a more appropriate term, given that an increase in occupant comfort range is usually subconscious. People are less likely to notice a deterioration in temperature and air quality when they are otherwise satisfied with a building's design.[11]

IT IS IMPORTANT TO HAVE BOTH MEASURED AND PERCEIVED PERFORMANCE TO GET AN ACCURATE UNDERSTANDING OF UNDERLYING ISSUES.

Key areas for occupant feedback include:

- perception and satisfaction with the thermal, visual and auditory environment
- design, usability and maintenance of spaces and systems
- anecdotal feedback on space planning, perceived health and wellbeing, productivity, wayfinding, access and security.

For non-domestic buildings, early-stage consultation can be greatly strengthened by systematic surveys to establish a baseline understanding of expectations and needs at briefing stage, against which feedback can be compared once a building is completed (see note 2.01).

The way people use buildings dramatically influences their energy consumption,[13] yet occupant activities are largely overlooked in the regulatory drive for energy efficiency.[14] Behaviour is not accounted for in building regulation compliance models, and building physics calculations primarily target fabric performance rather than occupant engagement.

This is one reason why projects where both IEQ and occupant satisfaction data are available are rare. It is important to have both measured and perceived performance to get an accurate understanding of underlying issues. Where these both exist, they provide fascinating insights into the changing needs of occupants, and inspiration to design for these.

FIGURE 2.3

Thermal comfort diagram, adapted from the WELL Building Standard.[12]

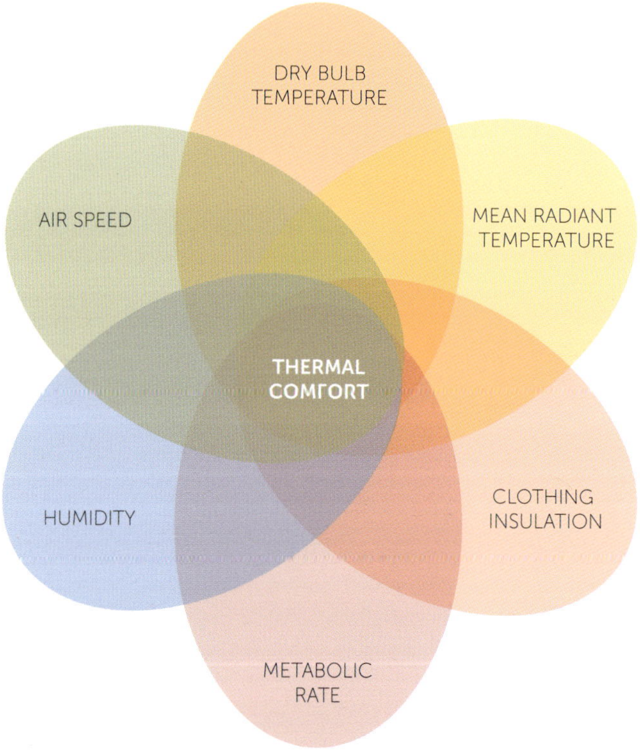

Thermal comfort

An important aspect of indoor environmental conditions is thermal comfort (see table 2.1 and figure 2.3), the perception of which affects occupant satisfaction. Thermal comfort is influenced by several factors: air temperature and velocity (draughts), temperature of surrounding surfaces (radiant temperature), and each individual's metabolic rate and clothing. Although every person's thermal comfort level is different, 80% of people will experience satisfaction within a similar range. Generally 23–28°C is considered a comfortable summer indoor temperature range, while 19–26°C is acceptable for winter indoor comfort.[15]

Until recently, the prevailing industry view among investors and policymakers[16] was that indoor comfort required a constant narrow range of environmental conditions. For several decades, the dominant thermal comfort model was based on the concept of thermal neutrality: i.e. the absence of discomfort from excessive heat, cold or humidity. People's comfort levels were measured under highly controlled constant conditions in thermal chambers, but this has in fact turned out not to be predictive of actual comfort levels in real-life conditions (see predicted mean vote/PMV).[17]

Instead, evidence has shown that the model of adaptive comfort is more reliable in predicting actual occupant comfort,[18] and can play a major role in building performance and occupant satisfaction. Building occupants like to be able to adjust internal environments in response to outdoor conditions, for example by opening windows, adjusting blinds and external shading, or turning up the heating locally[19] – see note 2.04. If occupants can adapt and make changes to their environment, much wider winter and summer comfort temperature ranges are possible. Adaptive comfort can play an important role in providing better thermal comfort with reduced energy use in both mixed-mode and air-conditioned buildings[20] – see notes 2.05 and 2.06 and also case studies 2 and 4.

Moreover, some degree of variability in the physical environment, such as changing lighting, temperatures, air movement and background noises, can positively affect the human body's circadian rhythm, mood and sleep patterns.[21] This relates to the concept of thermal alliesthesia, which uses the principle of 'thermal pleasure' that looks at the conditions required to create pleasure through encountering warmth when cold, and by being cooled when too hot.

Because thermal alliesthesia recognises subjectivity in thermal perception and the human body's need for variability, it challenges current industry practice in thermal comfort. Hence design thinking is moving away from the provision of static conditions as a panacea for comfort,[22]

2.04: EXAMPLE

INDIVIDUALLY CONTROLLED ADAPTIVE INTERVENTIONS

Thermal comfort is best supported by individually controlled adaptive interventions,[23] for example with the use of localised electric fans. For example in Taiwan 28°C indoor temperatures for moderate workloads are permissible.[24]

2.05: EXAMPLE

ADAPTIVE COMFORT IN HOT AND HUMID CLIMATES

Overcooling of buildings in hot and humid climates is frequently due to oversized system design and operation and is driven by legal and contractual requirements rather than occupant preference and comfort.[25]

2.06: EXAMPLE

OVERCOOLED SPACES ARE NOT PREFERRED

Global research involving 160 buildings demonstrated that occupants prefer wider temperature ranges than normally set in air-conditioned buildings, if these relate to outdoor conditions and they have control over their immediate environment.[26] Even in tropical climates, building occupants prefer slightly warmer indoor temperatures when combined with other measures such as reduced humidity or increased air flow to achieve thermal comfort, rather than 'overcooled' spaces. Temperatures as high as 29°C may be perceived as comfortable, if other adaptive measures are present.

2.07: EXAMPLE

VARIABLE CONDITIONS

Variation can be achieved by either changing conditions in the same space over time, or by moving through spaces with different conditions, such as short periods in transient spaces.[27] Brief exposure to higher (or cooler) temperatures, when moving from a comfortable zone, may be acceptable due to its short-term nature.[28] See also figure 2.4.

IF OCCUPANTS CAN ADAPT AND MAKE CHANGES TO THEIR ENVIRONMENT, MUCH WIDER WINTER AND SUMMER COMFORT TEMPERATURE RANGES ARE POSSIBLE and more towards biophilic design where indoor environments relate to the outdoor context[29] – see note 2.07 and figure 2.4. Moreover, a more holistic people-centred approach to thermal comfort is gaining ground due to the impact of the Living Building Challenge (see note 2.09), the WELL Building Standard and other initiatives.

The concept of thermal pleasure is likely to appeal to architects in particular because it offers greater design freedom in the provision of thermal comfort. Thermal alliesthesia advocates that dynamic environments enhance perception of thermal comfort. This can arise from changes in IEQ in response to outdoor conditions or equally from spending short periods in transient spaces.[30] Brief exposure to temperatures outside the comfort range may be acceptable.[31]

FIGURE 2.4
Instead of internal corridors, all circulation between classrooms takes place through the outdoor courtyard at Eddington Nursery in Cambridge (MUMA, 2018). This means that staff and students experience different thermal environments throughout the day as they move around the building.

Respiratory comfort

Respiratory comfort is another of the IEQ aspects that affects human comfort and satisfaction. It depends on the provision of fresh unpolluted air inside a building. Pollutants can come from external sources, such as traffic fumes and pollen, or from indoor sources, such as mould, particulate matter (PM) and volatile organic compounds (VOCs). The former requires filtering of the air intake, or moderating the times for opening windows. Indoor sources can be tackled at source, and this requires the careful specification of products, materials and finishes to limit chemical off-gassing. Increasing the air change rates will remove stale, humid air and fumes caused by human activities, but will increase energy use. Ventilation, which needs to be balanced with thermal comfort aspects, can also reduce humidity levels and help prevent mould growth.

The COVID-19 pandemic has acutely highlighted the importance of IEQ for health and wellbeing. The increased risk of disease contamination by building ventilation systems, many of which rely on recirculation of air to save energy, has led to calls to rethink the way in which fresh air and thermal comfort needs are met in buildings.[32] Sustainability and health concerns together highlight the need to find healthy and energy-efficient ways to provide more pollutant-free fresh air in buildings (see chapter 7).

Visual comfort

Visual comfort includes satisfaction with lighting conditions for different activities, specifically the amount, distribution and colour temperature of light (see note 2.08), as well as the absence of glare. With lighting comprising up to a third of energy use in non-domestic buildings,[33] daylighting provision is the most energy-efficient way to provide visual comfort. Lighting that is too intense or too low-level for a particular activity can affect concentration and productivity,[34] and a balance can be difficult to strike.

Maximising daylighting, without glare and undesirable (summer) heat gains is central to architectural design. The variability of the intensity, angle and colour temperature of light throughout the day animates spaces and textures. These transient qualities provide important stimuli for human wellbeing and are increasingly recognised for their role in regulating the body's circadian rhythms. Mimicking indoor daylighting with a uniform distribution of diffuse artificial light does not provide good visual comfort, based on feedback from buildings in use[35] – see case study 5 and note 6.02.

Visual comfort also depends on satisfaction with one's field of view. Views of the outside world, including the amount of sky visible from

2.08: EXAMPLE

ARCHITECTURAL QUALITIES AFFECT THERMAL COMFORT

Occupant thermal comfort is affected by architectural qualities such as building materials and finishes. For example in one study, increased thermal comfort was reported when participants were exposed to lower ('warmer') lighting colour temperatures. Participants reported feeling warmer when the space was lit with a warm (yellow-red) LED versus a cool (blue) LED, which made them put on more clothing. This early chamber-based research may indicate that tuning light colours could help reduce energy use in buildings.[36] However, these architectural qualities are not sufficiently quantified to capture thermal comfort or energy models.

2.09: EXAMPLE

LIVING BUILDING CHALLENGE

The Living Building Challenge (LBC) incorporates indoor environmental quality and criteria such as air quality and thermal and visual comfort in its Health + Happiness category. Operable windows for fresh air and daylighting and connection to nature are explicitly mentioned, as are occupant acceptance and engagement. The LBC also requires construction and building management and maintenance standards that support good IAQ, validated by post-occupancy IAQ tests.[37]

WITH LIGHTING COMPRISING UP TO A THIRD OF ENERGY USE IN NON-DOMESTIC BUILDINGS, DAYLIGHTING PROVISION IS THE MOST ENERGY-EFFICIENT WAY TO PROVIDE VISUAL COMFORT.

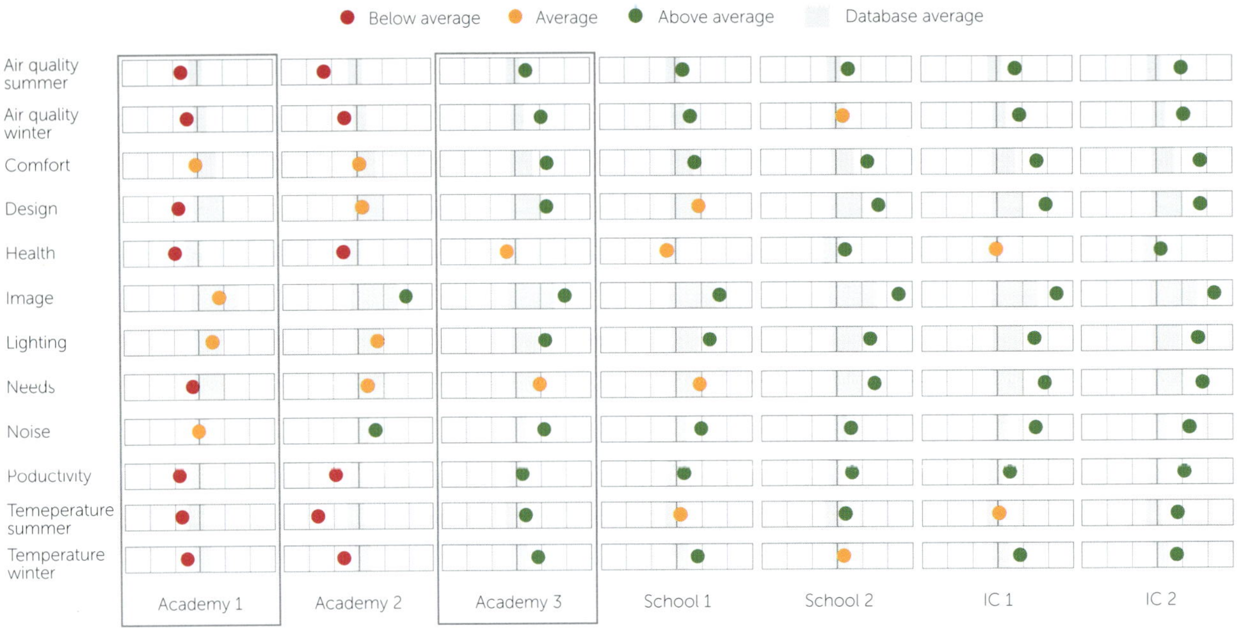

FIGURE 2.5
Summary of BUS survey results for 7 buildings studied as part of AHR/UCL Innovate UK building performance evaluations for academies 1 and 3, indicating the perception of occupant comfort and satisfaction. Differences in perception were not supported by IEQ measurements, which were similar in both schools, apart from acoustic conditions in school 1.[38]

indoors, and views of greenery and life, are important determinants of visual comfort. Perception of visual comfort also relies on the visibility of spaces in terms of privacy, and passive supervision of nearby spaces. New software tools are emerging to help designers analyse their proposals in terms of passive supervision, visibility and the movement of people.[39]

Acoustic comfort

Noise is sound that is perceived as too loud, and exposure to noise can have negative consequences for concentration, sleep, learning, productivity and mental wellbeing, clearly affecting human comfort and satisfaction. Noise comes from external sources such as traffic, or from internal sources such as voices or equipment in adjacent spaces. Internal noise exposure can be controlled individually, but often this is not possible, for example in open-plan offices. Feedback studies have shown that up to 60% of workers may be dissatisfied with acoustic interference in open-plan office spaces.[40] While acoustic conditions only require energy input for the operation of audio-visual equipment, blocking out unwanted noise is a major factor in the ventilation strategy of buildings (see chapter 6), and can incur higher energy consumption.[41]

Noise pollution can also have a detrimental effect on the perception of other comfort parameters, demonstrated by the example of two academy schools,[42] which had similar IEQ measurements yet diverging occupant satisfaction ratings (see note 2.10 and figure 2.5). This highlights the importance of combining both qualitative and quantitative feedback to analyse information and undertake appropriate remedial measures.

The role of benchmarks

A benchmark is a standard, parameter or performance target against which in-use feedback can be compared. It can also be a project database to compare performance against. Specific benchmarks related to human comfort and satisfaction are usually absent in sustainability certification frameworks, though undertaking occupant satisfaction surveys tends to be rewarded.

However, the Building Use Studies (BUS) methodology is a widely recognised benchmarking tool, referenced by BREEAM, LEED, the WELL Building Standard, NABERS and Soft Landings. The BUS Methodology was originally developed by Bill Bordass and Adrian Leaman through the Usable Buildings Trust in the 1990s.[43] It is a licensed procedure for undertaking standardised occupancy surveys that can then be used to benchmark a project against similar buildings in its database (see for example figure 2.5). The top-level BUS benchmarking parameters include:

- temperature in summer/winter
- air quality in summer/winter
- overall comfort
- design
- health
- image
- lighting
- needs
- noise
- productivity.

The UK Chartered Institute of Building Services Engineers (CIBSE) has also issued benchmarks and guidance on indoor environmental quality that can be found in, for example, Technical Memoranda (TM) 40, 52 and 59, and energy benchmarks in TM54[44] and in the London Energy Transformation Initiative (LETI) Climate Emergency Design Guide.[45] The RIBA adopted CIBSE and WHO benchmarks for its Climate Challenge targets.[46] Benchmarks, are vital tools for all built environment professionals to understand the performance of their buildings. To glean a thorough understanding of the human factor in building design, quantitative indicators and qualitative input are both needed. Used in tandem with benchmarks, this information provides powerful insights into how buildings work in use, and where they are on the journey to net zero.

2.10: EXAMPLE

IMPACT OF NOISE POLLUTION ON OTHER COMFORT PARAMETERS

Two academy schools in identical weather zones, with similar indoor comfort parameters in terms of temperature and CO_2 measurements, had diverging comfort perceptions. The occupant survey of academy 1 found high degrees of dissatisfaction with acoustic conditions – the building's acoustic properties amplified the noise from open-plan learning, interfering with teaching. As a result, the satisfaction levels across all other IEQ indicators suffered, including thermal comfort and air quality, despite these being well within the comfort range. This contrasts with nearby academy 3, with similar IEQ measurements, and no obvious differentiating aspects from School 1 apart from having no acoustic challenges. In this case, the building met occupant needs well and they responded positively to all IEQ parameters. See figure 2.5.[47]

A BENCHMARK IS A STANDARD, PARAMETER OR PERFORMANCE TARGET AGAINST WHICH IN-USE FEEDBACK CAN BE COMPARED

03

ENERGY BASICS AND BENCHMARKS

FIGURE 3.1: OPPOSITE
Keynsham Civic Centre, Keynsham, AHR.

Introduction

Energy is used in buildings to supplement a building's passive provision of fresh air, heat, coolth, light and functionality. Energy use can be a useful proxy for building quality, as it provides an insight into how much 'effort' it takes to support human comfort and productivity. As such, measuring energy use helps to identify wasteful practices in design, construction and operation which impact on occupant comfort as well as maintenance and operating costs. Tracking energy use alongside indoor environmental quality and occupant satisfaction provides a more robust approach to building performance.

How much energy is 'too much' depends on what it is compared to, and benchmarks can serve as a useful baseline. Good benchmarks depend on high-quality data. Unfortunately this data is not yet readily available, because in most countries it is not usually a requirement to report the operational energy use of buildings.

Nevertheless, it is agreed that all building projects should be targeting 'zero energy' use to avoid catastrophic heating of the planet. This is why understanding where energy is used and reporting energy consumption in a harmonised manner is essential.

Energy use metrics

Energy consumption metrics vary from design to operation. Understanding how they stack up is essential to be able to set targets and compare these with measured consumption. The following energy literacy terms will not only help you navigate this book, but are essential knowledge for anyone involved in the building industry.

(Net) zero energy

The concept of 'net zero' energy can be confusing, with different terminology in circulation – see note 3.01. Put simply, it means that a building generates as much on-site energy as it uses. If a building's CO_2 emissions are (net) zero carbon, this means that renewable energy is being used, generated either on or off site, or that carbon emissions are being offset. Some definitions refer to net zero 'operational' carbon,[1] while others refer to net zero whole-life carbon emissions,[2] which also includes embodied carbon.

Carbon vs energy

For many years experts have debated whether to report energy performance in terms of carbon emissions or energy consumed. Carbon remains the high-level metric to account for a building's climate change impacts. However, metrics can have unintended consequences. For example, if carbon is used as a design goal, a building can still be wasteful of energy as long as its demand is met from renewable sources (see note 3.01). The greater the energy consumption, the more difficult it is to meet all energy needs from low-carbon sources. Moreover, zero carbon technologies also have a limited life span in a building. Low-energy buildings are more robust and able to withstand the eventual failure and replacement of low-carbon technologies, without significantly increasing carbon emissions. Reducing energy use remains the most important goal.

Energy intensity

Energy intensity (kWh/m²/yr) is the accepted reporting metric for building energy use. However, measuring and reporting energy use per square metre or per person or per cubic metre can make the same consumption appear more efficient. For example, reporting per square metre does not capture occupant density, as more people and longer operating hours can mean more energy use per square metre.[3] This is why the commercial sector usually prefers to report consumption per person. However, the impact of occupant density is unclear (for example, flexible working might leave allocated desks empty), and person-hours of occupation might be a better additional indicator along with kWh/m²/yr.[4] The impact of occupant density on energy use could be offset in an energy-efficient building, for example by adjusting occupancy patterns and with good energy management practices.[5] See also note 3.02.

Carbon intensity

Electricity and heating fuel consumption are converted to carbon emissions using 'carbon intensity factors' ($kgCO_2/kWh$) that are set nationally. They

3.01: DEFINITION

NET ZERO ENERGY, NET ZERO CARBON, ZERO ENERGY, ZERO CARBON, CARBON NEUTRAL OR NEARLY ZERO

These region-specific terminologies can all mean something slightly different, depending on whether a building uses fossil fuels and grid electricity at all, and how much of the energy used is generated on site. A building can be zero carbon but this does not automatically mean it is net zero energy, or even energy efficient.

3.02: EXAMPLE

VOLUMETRIC ENERGY REPORTING

In countries such as Italy, where historic buildings with tall floor-to-ceiling heights form the majority of the building stock, energy use is often reported per cubic metre to help make the existing stock seem comparable in terms of efficiency to new buildings.

THE GREATER THE ENERGY CONSUMPTION, THE MORE DIFFICULT IT IS TO MEET ALL ENERGY NEEDS FROM LOW-CARBON SOURCES.

ENERGY INTENSITY (KWH/M²/YR) IS THE COMMON REPORTING METRIC FOR BUILDING ENERGY USE.

FIGURE 3.2

When modelled, building energy use is often referred to as 'by end use', or when measured in terms of the energy sources used ' by fuel', both of which are converted to CO_2 emissions using carbon intensity factors.

3.03: EXAMPLE

SUBMETERING BY LOCATION

Certain reporting frameworks, such as the UK's Design for Performance, require submetering by location so that the performance of landlord areas can be benchmarked separately from tenant consumption.[6] Sub-wiring by space can be costly for multiple tenants in large buildings, but worthwhile where there are multiple occupants impacting on building performance.

3.04: EXAMPLE

TYPICAL FUEL SOURCES OF END USES

Heating and hot water is traditionally supplied by burning fossil fuels on site, while grid electricity supplies all other end uses. But other technologies exist that use heat to generate electricity on site (e.g. combined heat and power/CHP), or use electricity to supply low-carbon heating or cooling (e.g. heat pumps). This adds a level of complexity and requires elaborate metering to enable 'by fuel' consumption with energy 'end uses'.

vary depending on the CO_2 intensity of the local energy generation (i.e., the amount of CO_2 emissions released per kWh of fuel used). This means that two buildings which consume the same amount of energy can be low carbon in one country but carbon intensive in another, if more or less of the national energy is generated from zero carbon sources.

Energy use 'by fuel' or 'by end use'

Energy in buildings is consumed by technical systems and equipment. Energy consumption can be discussed in terms of what it is used for, often referred to as 'by end use', or in terms of the energy sources used, i.e. 'by fuel'. The calculated output from a design-stage thermal model is energy consumption 'by end use', while mains energy use is typically measured in buildings 'by fuel' in terms of electricity, heat (mostly gas) or clean energy consumption – see figure 3.2. Energy data 'by fuel' helps with understanding the scale of energy consumption. Energy 'by end use' data helps us understand in more detail where and why there might be underperformance, and how this can be improved and avoided in the future – especially if accompanied by a predicted baseline – see figure 3.3. and also note 3.04.

Meter reconciliation

While mains meters measure 'by fuel' energy consumption, submeters measure energy end uses. The consumption measured by all submeters must add up to the total measured by the mains meters. This exercise is called 'meter reconciliation', and it is a critical element in commissioning building systems and validating predicted performance in use (for more, see chapter 8).

Measuring energy end uses in a building and comparing them with predicted ones is the first step for building energy use diagnostics as described by the UK Chartered Institution of Building Services Engineers (CIBSE) Technical Memoranda *TM22: Energy Assessment and Reporting Methodology*.[7] See also note 3.03.

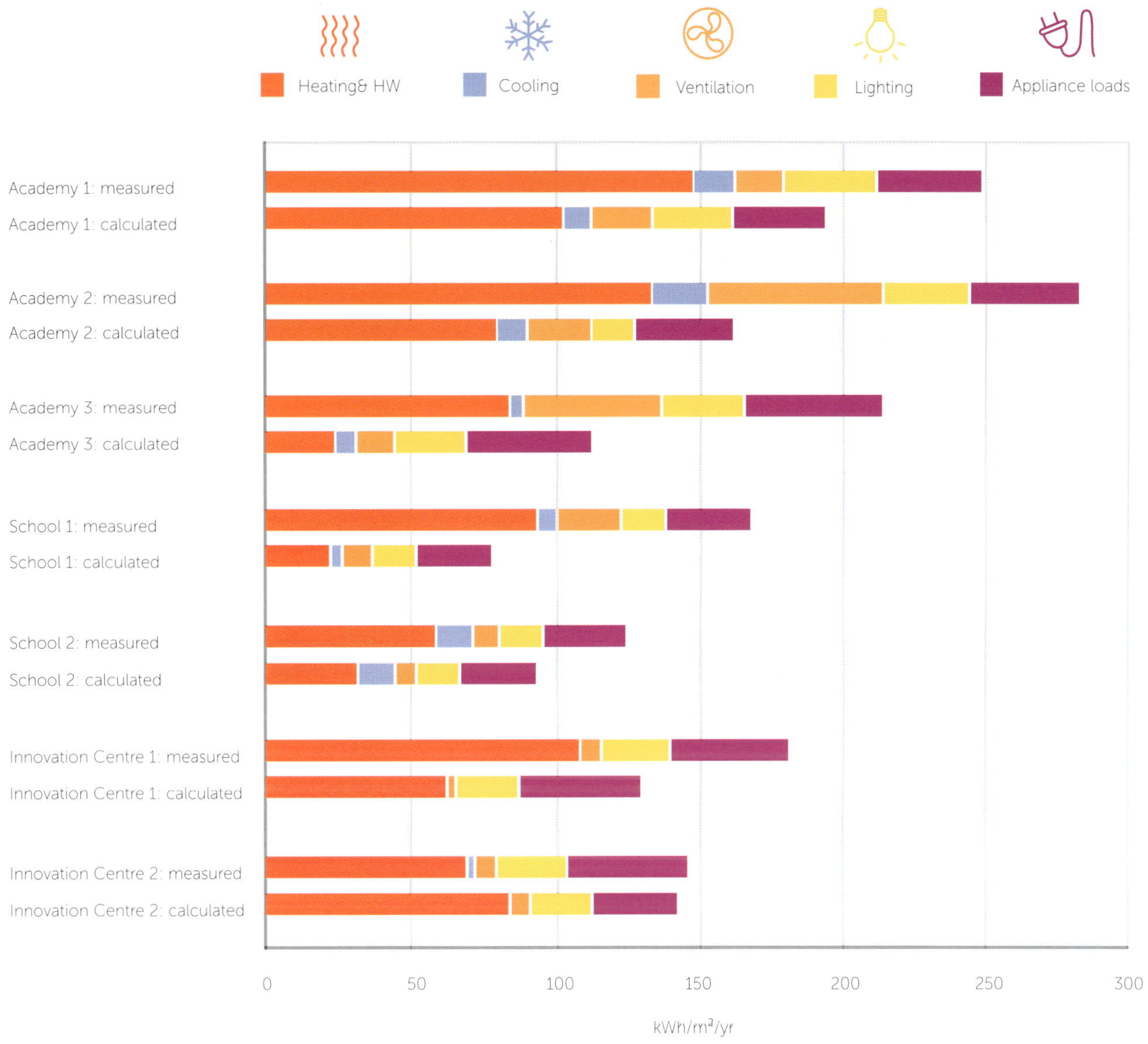

FIGURE 3.3

Energy bar diagram comparing the calculated and measured energy use of seven buildings monitored by AHR and UCL architects. The visual energy bars clearly indicate that, in most cases, measured energy end use dramatically exceeds predictions.

Communicating measured building energy use

When communicating 'by fuel' energy use, it is controversial to place measured kWh of heat energy side by side with kWh of electricity consumed on an 'energy bar'. This is because the same heat kWhs are not equivalent to kWhs from electrical energy use, due to transmission loss in the electrical grid, with as much as 70% of the energy being lost in conversion and transmission.[8] To make both fuel uses comparable, they must be translated into primary energy, which includes these transmission and conversion losses. Nevertheless, it is common practice to communicate total kWh energy use by fuel on the same horizontal bar – see figure 3.2. Be aware that often benchmarks give total energy kWh/m^2/yr without conversion to primary energy.

Daily energy profiles

Daily energy use profiles are helpful to diagnose issues with mechanical services and can be measured using smart meters or building management systems (BMS) – see chapter 8. For example when a building is unoccupied at night and in the early morning, technical systems should reduce their outputs and hence reduce energy use. As figure 3.4. indicates, this did not happen in academy 2, which was diagnosed by its daily electricity profile and higher-than-expected energy use.

Tracking energy use metrics from design to operation provides a powerful mechanism to ensure that project goals set at inception are achieved in practice. See also chapter 10.

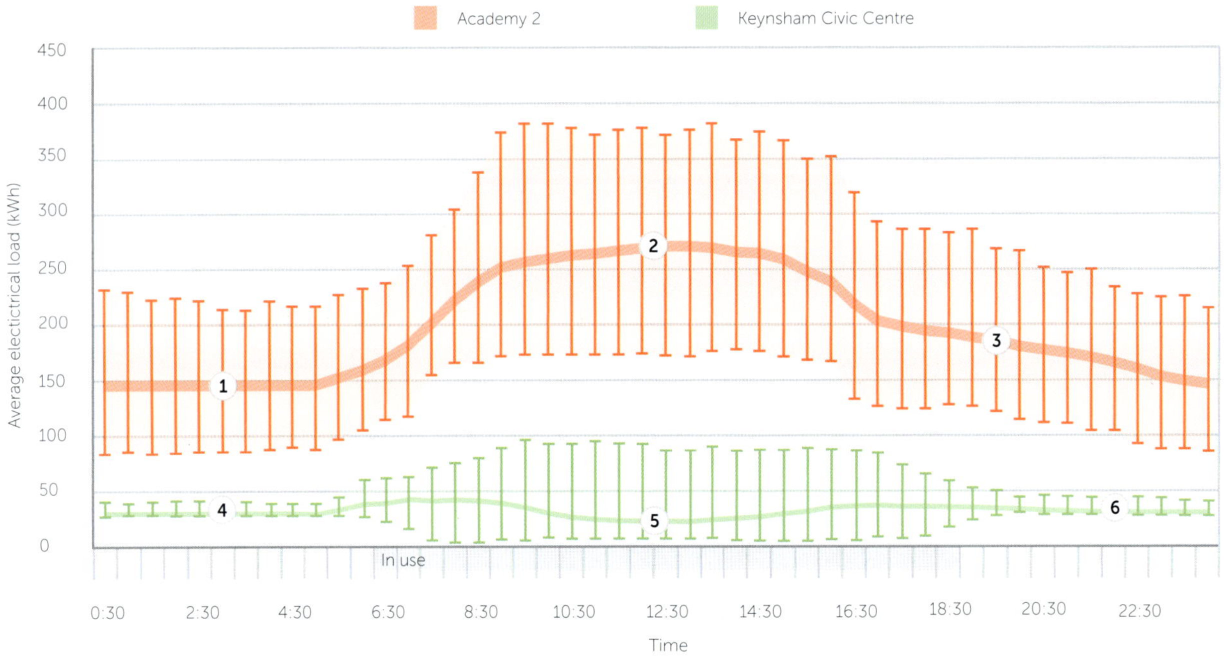

FIGURE 3.4

Daily electricity profiles at two buildings show a high energy baseload (orange) throughout the day for academy 2, with only a slight drop after school hours, and a lower baseload (green) and out-of-hours use of the more efficient Keynsham Civic Centre (case study 6).

1 Base load when building is empty

2 Peak load

3 Out of hours occupancy affecting whole building base load

4 Base load of more energy efficient building

5 Peak loads offset by pv generation

6 No significant impact from out of hours occupancy

Benchmarking

Benchmarks are standards or points of reference. They can be used to set building energy use targets at the design stage, to compare to post-completion measured performance. Good granularity (or detail) of data is important, so comparisons are meaningful.

When comparing energy benchmarks, it is important that they are recent, ideally of similar building types and build standard, and that they refer to measured consumption. For example, a newly constructed building's energy use should not be compared to a building constructed to different regulations decades ago. As shown in table 3.1, benchmarks can be based on actual case study buildings, or on theoretical building energy models. They can also be based on statistical analysis of the energy consumption of a large number of buildings that include as much detail as possible on building energy determinants, such as plan layout, fabric, systems and operating conditions. For example, the Energy Star benchmark[9] is an American statistical tool with demonstrated methodology using a combination of measured data and statistical calculations to arrive at robust benchmarks.

At early design stages, statistical data can be a useful initial predictor of expected energy use, but will not be accurate. Lack of high-quality public data worldwide means that benchmarks are often out of date, which undermines confidence in their use. This is why many researchers and professionals across the globe are campaigning for better reporting practices, to improve benchmarking based on performance data and support better regulatory and procurement practices.[10]

ECON19, a benchmark for different types of office buildings published by the Usable Buildings Trust almost two decades ago, has been widely used in the UK because it includes a breakdown of energy end uses.[11] In the absence of more recent data with this high level of detail, the online CarbonBuzz benchmarking project gathered detailed energy 'by fuel' and 'by end use' data from design to operation of different non-domestic building types based on actual performance.[12] Using crowdsourced data, the CarbonBuzz project helped expose the scale of the performance gap between the calculated and measured performance of buildings.

One of the more comprehensive UK heat and electricity consumption benchmark databases is CIBSE technical *Guide F*, which provides energy use profiles for typical building types based on data gathered from public buildings since 2006 through the UK's Display Energy Certification (DEC) programme – see note 3.05.[13]

In the absence of sufficiently detailed benchmarks, building regulations rely on compliance calculations to benchmark energy performance at design stage. However, neither of these 'benchmarks' are a suitable proxy for predicting actual operational energy use.

TRACKING ENERGY USE METRICS FROM DESIGN TO OPERATION PROVIDES A POWERFUL MECHANISM TO ENSURE THAT PROJECT GOALS SET AT INCEPTION ARE ACHIEVED IN PRACTICE.

3.05: DEFINITION

DISPLAY ENERGY CERTIFICATION (DEC)

Display Energy Certification (DEC) is the methodology for benchmarking in-use energy consumption for non-domestic buildings in the UK. A DEC must be displayed in all public buildings with an area larger than 500 m^2 after one year of operation, and updated annually. DECs and similar frameworks in other countries offer a helpful reporting framework for setting operational energy targets for non-domestic buildings, as long as up-to-date carbon intensity factors are used. However, fuel CO_2 intensity factors become rapidly out of date when countries increase their renewable energy capacity.

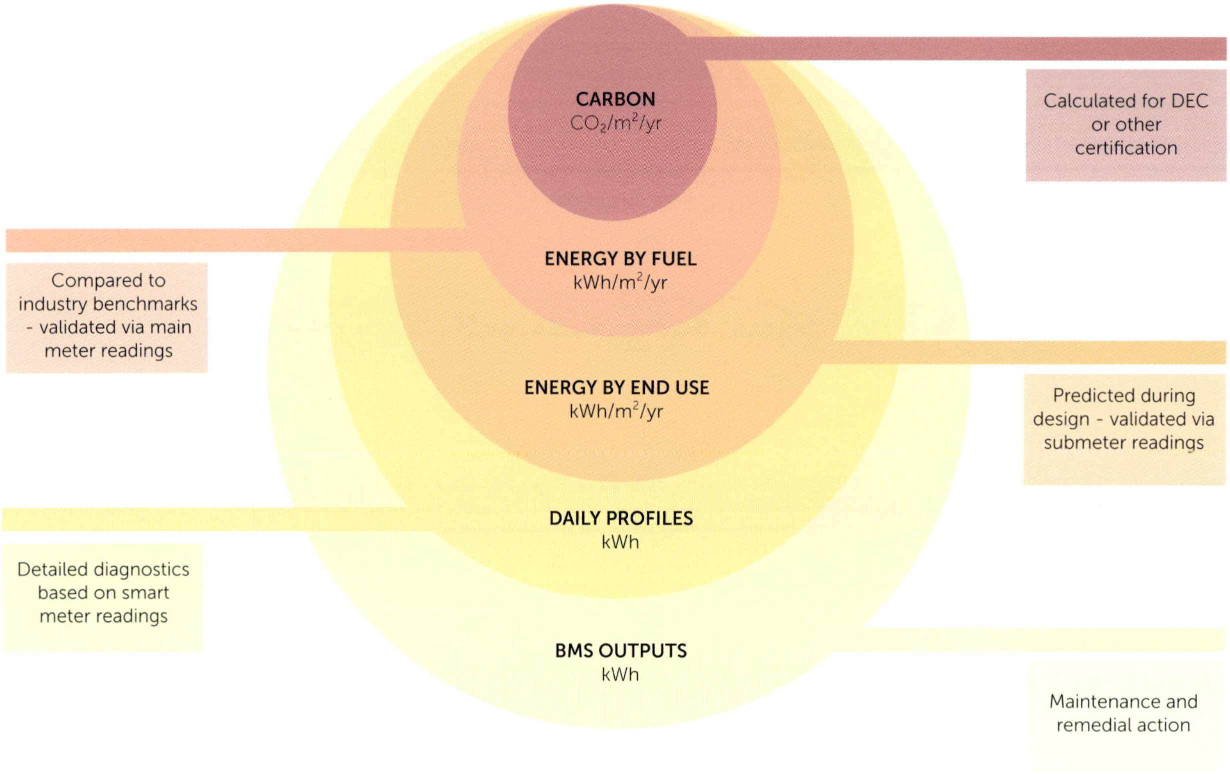

FIGURE 3.5

This diagram summarises the different levels of detail of energy data and explains how they are used.

Operational energy frameworks and standards

In addition to energy benchmarks, there are frameworks that specifically target operational energy use. This includes BSRIA Soft Landings and the Better Buildings Partnership's Design for Performance scheme.[14] The latter was developed for multi-tenanted buildings and targets the base building performance only – heating, hot water, ventilation and air-conditioning of the whole building, light and power in common areas, and the lifts, as these are controlled by the landlord.[15]

At the time of writing there are no official in-use benchmarks for operational energy use in housing in the UK. There are also data privacy issues relating to the collection of energy data for homes, making statistical benchmarking less effective for private dwellings.

One particularly tricky aspect of feedback processes is accurately calculating a building's energy performance. Different methods for calculating and predicting performance are explained in the following chapter.

TABLE 3.1: THE MOST COMMONLY USED BENCHMARKS AND THEIR SCOPE

CIBSE TM46 (2008)	Median electric and non-electric energy use for 29 benchmark categories covering 237 building types. DECs, also referred to as 'operational ratings', are based on these and inform other best-practice schemes.
Display Energy Certification (DEC) (2006)	DECs use 2006 carbon factors to convert CIBSE TM46 energy benchmarks to carbon emissions that are translated to A–G ratings. The D rating is derived from the TM46 benchmark for a particular building type, allowing adjustment for mix of use, operating hours and special energy uses. See also note 3.05.
CIBSE Guide F[16] (2012/2016)	CIBSE Guide F describes typical and good-practice fossil and electricity consumption benchmarks for 92 use types. This is accompanied by an online benchmarking tool that allows users to compare a project to a distribution of projects using a combination of CIBSE Guide F and TM46 benchmarks, the DEC register, and other databases where data sharing has been agreed. In future, users will be able to contribute to the benchmarks by uploading data.
LETI Climate Emergency Design Guide (2020)[17]	The London Energy Transformation Network published recommended energy use intensity benchmarks for housing, commercial offices and schools to achieve zero carbon buildings by 2030.
RIBA 2030[18] Climate Challenge (2019)	The RIBA 2030 target for new non-domestic buildings of 55 kWh/m^2/yr is based on the energy use intensity equivalent of a DEC A rating for an office building, and aligns with UKGBC targets for zero carbon offices.[19] The target for new housing of 35 kWh/m^2/yr is based on a 70% reduction from the OFGEM benchmark. Both align with LETI guidance.
CarbonBuzz (2005)[20]	Online benchmarking tool primarily for non-domestic buildings, offering interactive comparisons with data from anonymised crowdsourced projects as well as published portfolios, including the UKRI BPE programme. Both energy consumption 'by fuel' and, uniquely, higher granularity data 'by end use' is recorded.
Passivhaus	Passivhaus is a certification programme that focuses on operational energy use reduction, setting maximum thresholds for space heating and cooling demand, airtightness and overheating. It is a framework for achieving high construction quality, especially with regard to airtightness and fabric insulation in reality.[21]
Energy Star	The US Energy Star programme recognises energy consumption below a maximum threshold for electrical and non-electrical energy use.
NABERS	Australian benchmarking and rating tool for office buildings, rating performance based on electrical and non-electrical energy use for landlord areas only, excluding tenant space. Provided the basis for the UK Design for Performance initiative led by the Better Buildings Partnership (BBP).[22]

RIBA 2030 CLIMATE CHALLENGE SETS A TRAJECTORY TOWARDS NET ZERO PERFORMANCE FOR PROJECTS AND INCLUDES EXPLICIT KWH/M2/YR TARGETS FOR OPERATIONAL ENERGY USE FOR DOMESTIC AND NON-DOMESTIC PROJECTS. SEE FIGURE 11.2 TO COMPARE CASE STUDY ENERGY USE INTENSITY TO RIBA 2030 TARGETS.

04

DESIGN FOR FEEDBACK

FIGURE 4.1: OPPOSITE
Carrowbreck Meadow, Hellesdon, Norwich.

Introduction

Feedback makes explicit – and helps quantify – the value of good architecture. This is because building performance can be used as a mechanism to increase accountability of all parties throughout design and construction, particularly if it is embedded in procurement processes, as outlined in chapter 10. It empowers architects to advocate for the long-term needs of building users, and for public benefits.

Performance evaluations allow all parties – clients, designers and contractors – to learn from past projects and reduce the risk of underperformance across future projects.[1] Feedback has reputational benefits, too. By sharing evidence and knowledge publicly, organisations can demonstrate thought leadership in addressing the global climate challenge while gaining professional recognition for their skills and expertise. In a departure from the industry's tendency to bury embarrassing underperformance, this constructive and open approach is slowly galvanising professionals and educators around the world.[2]

Despite the compelling arguments for feedback, substantial barriers have prevented more widespread take-up of feedback practices. This chapter reviews the challenges to feedback and sets out a pathway for its more widespread adoption.

The feedback process

Clarity at the outset of a project about the purpose of feedback helps determine what type of performance studies should be undertaken, and at which stages in a project's life cycle. Post-occupancy evaluations (POEs) are carried out post-construction, whereas building performance evaluations (BPEs) prescribe activities across all project stages, and conclude with the comparison of measured performance in use with design stage calculations and benchmarks. While any feedback is useful, a BPE approach throughout all project stages offers maximum benefit.

To get the most out of BPEs, design teams must simultaneously agree on indicators or metrics for quantitative analysis, and also record qualitative parameters to help contextualise metrics (see chapter 2). As outlined in the preceding chapters, key parameters for feedback collection include occupant satisfaction, indoor environmental quality (IEQ) and energy consumption. To scrutinise whether key performance criteria have been achieved, key performance indicators (KPIs) can be used. Clarity is key, and it is paramount not only to define the objectives of feedback, but to secure 'buy-in' from the entire project team. More detail does not necessarily lead to greater insights or better in-use performance. Instead, when gathering feedback from buildings in use, a 'graduated response approach'[3] is optimal, where high-level indicators are monitored and greater granularity is only sought as and when necessary to explain any anomalies.

CLARITY AT THE OUTSET OF A PROJECT ABOUT THE PURPOSE OF FEEDBACK HELPS DETERMINE WHAT TYPE OF PERFORMANCE STUDIES SHOULD BE UNDERTAKEN, AND AT WHICH STAGES IN A PROJECT'S LIFE CYCLE.

Starting early

While evaluations can be helpful even if only carried out in hindsight, where a feedback process is planned from the outset, benefits are magnified because design assumptions and targets can be tracked throughout a project's duration. This means that data is more readily available and expectations are managed. Incorporating targets into building contracts and consultants' scopes further increases accountability. Ideally, a mechanism for rectifying any problems that emerge post-completion should be agreed early on. This comprehensive approach to feedback results in better buildings and exponential gains in design team productivity, where everyone works towards the same goal.

Using feedback from similar past projects, or carrying out a pre-project evaluation of existing premises, gives designers a more factual understanding of a brief, rather than relying on intuition. Feedback also improves project outcomes, because it counteracts professionals' tendency to work in isolation by incentivising the sharing of information and collaborative problem solving (see case studies 4 and 7). Problems that inevitably arise post-completion can be resolved more promptly, demonstrating value for money to clients.

USING FEEDBACK FROM SIMILAR PAST PROJECTS, OR CARRYING OUT A PRE-PROJECT EVALUATION OF EXISTING PREMISES, GIVES DESIGNERS A MORE FACTUAL UNDERSTANDING OF A BRIEF, RATHER THAN RELYING ON INTUITION.

Finally, both a willingness to tackle project delivery collaboratively and the demonstration of client and occupant satisfaction with past projects will lend credibility in competitive tenders, and can help architects win work.

Calculating a performance baseline

The challenge of accurately predicting baseline energy performance that can then be monitored across different project stages is a complex topic that lies at the heart of the performance gap – see figure 4.2. Numerous tools and standards have been developed for predicting a baseline building energy use. These may be based on climate, context, function or building characteristics, as well as occupancy and operating patterns. It is important to distinguish between compliance calculation methods used to demonstrate compliance with building regulations (in the UK: SAP or SBEM) and certification (e.g. Energy Performance Certification/EPC). There are also building system and services models used by building service engineers to design mechanical systems, and those used for predicting actual baseline energy consumption of buildings in use.

Confusion arises because compliance/certification models are often mistakenly interpreted as predictive calculations, when in fact such

FIGURE 4.2

Performance gap diagram, showing what is typically included in calculated energy models compared to actual measured energy consumption.

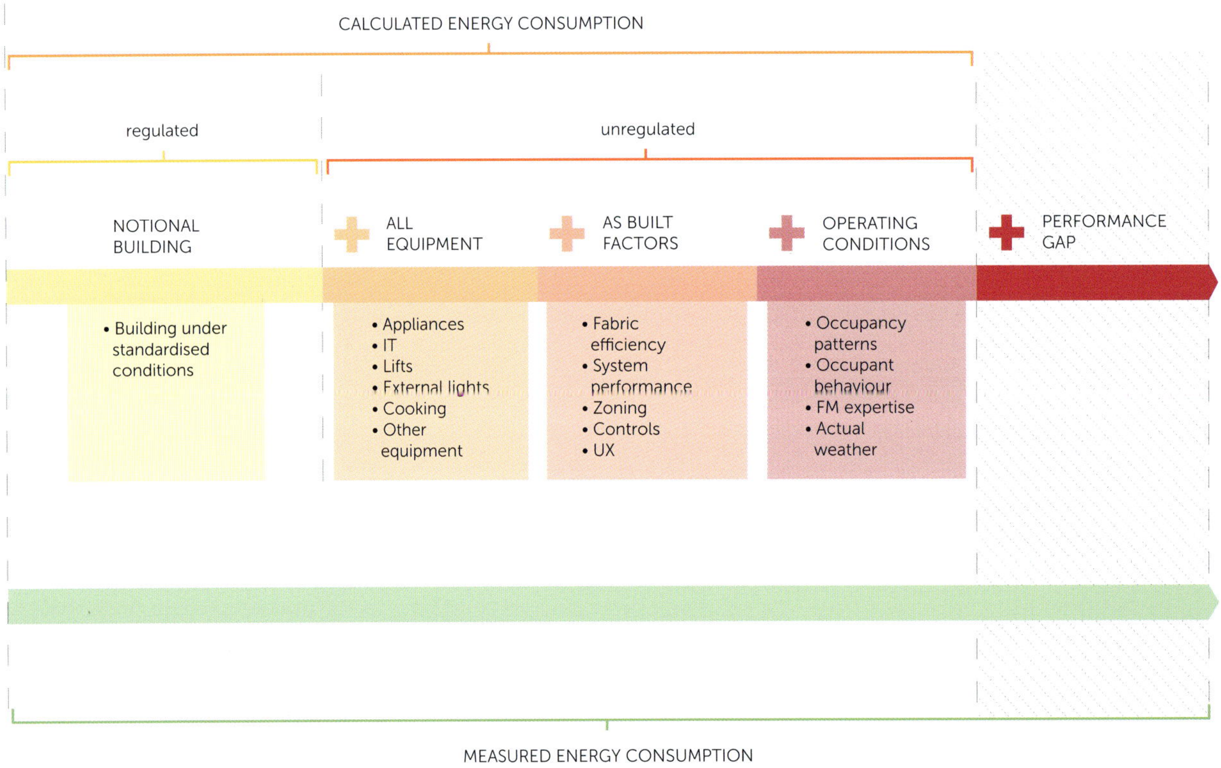

models were never intended for this purpose. They assume standardised theoretical conditions (sometimes of notional buildings) to enable comparison between different buildings. This means that critical variables that can dramatically impact a building's resource consumption, such as actual operating conditions, building fabric, systems and spatial configurations, and electrical plug-in loads, are completely ignored. For compliance calculations, engineers often use steady state models, which are less precise because they do not reflect daily and hourly variations in demand.

Building system and services models_involve more elaborate tools that are used for system and services design by engineers. These are often dynamic energy models that, unlike steady-state models, use hourly weather data to calculate a building's response to varying external and internal conditions. Typically, dynamic models include assumptions about operational conditions such as building management, natural ventilation practices, and system efficiencies and their usage, but not necessarily in a realistic way. These models often ignore occupancy patterns, or are based on unrealistic assumptions.

The discrepancies between in-use energy consumption and the predictions of both compliance and building system and services design models contribute to the performance gap. Yet, both models could be adjusted to more accurately predict a building's energy consumption (figure 4.2). For example, compliance models could be refined to include more detailed inputs, and system design models can be expanded to include operational inputs based on the client's actual operational management practices. However, this does not usually happen in practice. The point is that every single energy use must be 'counted' and minimised.

Moreover, architects and clients rarely see the inputs and assumptions that these models are based on, because this disclosure could present a professional indemnity (PI) insurance risk to engineers by opening project assumptions to scrutiny and interpretation. For example, if model assumptions prove unrealistic once the building is completed, engineers might be held responsible under the 'due diligence' clause, unless the liability terms are clearly set out in contractual clauses. Another reason why the full project team is rarely involved in agreeing the assumptions that underpin these predictive models is that it is common practice to charge extra fees when services design models are used for actual building energy use estimates. Few clients are willing to pay for this additional service.

Energy modelling for the future – a step-change

It is clear from the above that the current approach to predictive building energy use models requires a reality check. The same advanced dynamic models used for building system and services design could be adapted to predict actual energy consumption for all the typical regulated energy 'end uses', such as heating, cooling, lighting and ventilation (which are typically regulated for in building legislation), and also the unregulated auxiliary power and electrical appliance plug loads (usually excluded and unregulated by building legislation) – see figure 4.2. If accurate building properties and operating conditions are identified and coordinated with the client and the design team, predicted energy use can be within 5–10% of actual energy instead of 30% plus.[4] The use of feedback, both for a specific building and through systematically monitoring buildings in use, is a robust way to identify realistic model assumptions to bridge the performance gap.

Barriers to feedback

Despite the compelling case for feedback, there are numerous barriers to its holistic incorporation into building practices, such as the predictive energy modelling issues described earlier. Additional barriers are discussed in this section.

Professional indemnity (PI) insurance

In the United Kingdom, the prevailing advice of PI insurers has been not to carry out any post-completion performance evaluation. Gathering feedback may detect adverse performance and satisfaction issues, which, short of significant failure, would otherwise remain invisible. Design teams might be uncomfortable with such exposure, fearing reputational damage or a hike in PI premiums.

However, this situation is changing, and decisions are now made on a case-by-case basis. Some insurers will cover the risks associated with post-completion evaluations as long as they are notified in advance, and they generally advise that such work be carried out under a separate contract. This is so that the insurance period for project design and construction (which can be up to a decade or more) ends on practical completion. Instead, feedback processes, which generally take between one and three years post-completion, carry a much lower level of risk.

The most frequent defects uncovered by feedback relate to thermal bridging, metering and control systems. These are defects which, if caught early, can be fixed relatively cheaply and bring major benefits in terms of the satisfaction and comfort of occupants, and long-term maintenance and operational costs.

Evidence shows that legal action is less likely to occur when there is continued dialogue, even about major problems. It is rare for performance evaluations to trigger litigation – and where it does happen, the issues would have most likely emerged regardless, and in this way they are resolved more quickly.[5] At the UEA Enterprise Centre (case study 7), project team collaboration through the Soft Landings process prevented a relatively minor building management system (BMS) malfunction from escalating into a potentially litigious issue. Rather than reinforcing a culture of blame, performance evaluations have been shown to reduce risks by promoting a culture of shared accountability.

RATHER THAN REINFORCING A CULTURE OF BLAME, PERFORMANCE EVALUATIONS HAVE BEEN SHOWN TO REDUCE RISKS BY PROMOTING A CULTURE OF SHARED ACCOUNTABILITY.

Procurement

Design and build (D&B) contracts pose a particular challenge for feedback studies (see also chapter 1). If design team members undertake feedback as part of a D&B contract, they could encounter a conflict of interest when they are novated to a main contractor and have to communicate potentially adverse findings to the client or end users. In these circumstances, use of an independent external evaluator is essential, and should be included as a requirement in the contractor's preliminaries.

In general, if the gathering and reporting of performance data is included in the contractor's scope, the design team's cost estimate for basic benchmarking exercises can amount to less than a third of current cost estimates, especially if carried out in-house by original team members.[6] With this approach, design teams gain the greatest knowledge from feedback processes. The costs can be viewed as continuous professional development (CPD), and in some countries feedback processes may qualify for research and development (R&D) tax relief.[7]

Independent verification of results by an external or academic evaluator is advantageous, because it inspires greater confidence in the findings. On the other hand, the alternative approach of having the design team or contractor responsible for post-completion evaluation enables continuity of project information, which helps keep evaluation costs low.

An evaluation team appointed post-completion can make recommendations for rectification of problems, but acting on these remains the responsibility of the client. It is crucial to assign authority to act on the findings. Setting aside a budget at the outset to address post-completion issues is an effective way of quickly resolving problems and avoiding conflict and blame. Avoiding adversarial contractual relationships encourages wider take-up of feedback processes.

Dramatic improvements in performance can be achieved when the requirement to provide feedback on specific performance indicators is incorporated within the contractor's preliminaries, effectively creating a performance contract, and establishing shared incentives. This is further discussed in chapter 10 (see case studies 4, 6 and 7).

DRAMATIC IMPROVEMENTS IN PERFORMANCE CAN BE ACHIEVED WHEN THE REQUIREMENT TO PROVIDE FEEDBACK ON SPECIFIC PERFORMANCE INDICATORS IS INCORPORATED WITHIN THE CONTRACTOR'S PRELIMINARIES.

Costs

The cost of a BPE is estimated to add between 0.1% and 0.25% of the upfront cost to a project when it has not been planned for from the outset – see figure 4.3.[8] This can amount to a significant sum on a large project, and helps to explain why feedback processes have been slow to gain traction. However, the cost of sub-optimal operation over a building's lifetime can dwarf the cost of undertaking feedback processes, which could highlight issues that can be rectified. Yet persuading clients to spend additional funds after completion, when everyone just wants to move in and move on, is difficult, no matter what the benefits may be. This is another compelling reason for establishing a BPE process upfront.

IF FEEDBACK LOOPS ARE PLANNED FOR FROM THE OUTSET, THE WHOLE PROCESS CAN BE STREAMLINED.

If the project team has already disbanded, as much as half the time can be spent gathering and reconciling data due to the likelihood of missing logbook entries and poorly set-up and non-reconciled meters and sensors. Another quarter of the time can be spent chasing clients, occupants and design team members for permissions, and to track down design information to support building diagnostics. That leaves just a quarter of the time left to interpret the data and make recommendations – see figure 4.3.

If feedback loops are planned for from the outset, the whole process can be streamlined.

When responsibilities are clearly allocated and proper metering is correctly installed, the flow of data can be near-automatic, even remote, dramatically cutting the cost of feedback evaluation. To ensure that the full benefits of feedback processes are realised, it is important to allocate project time and costs upfront so that sufficient budget is left after construction for feedback to be collated and improvements undertaken, as well as further monitoring to validate performance.

FIGURE 4.3

The cost of feedback: Diagram showing costs of feedback as a proportion of capital costs when planned for ahead (left) versus when initiated at project completion (right).[9] After the original project team is dissolved, acquiring design and as-built information and gathering energy data is time-consuming. When feedback is incorporated from the outset, costs are reduced dramatically. Moreover, underperforming systems can be rectified early, further saving costs for clients.

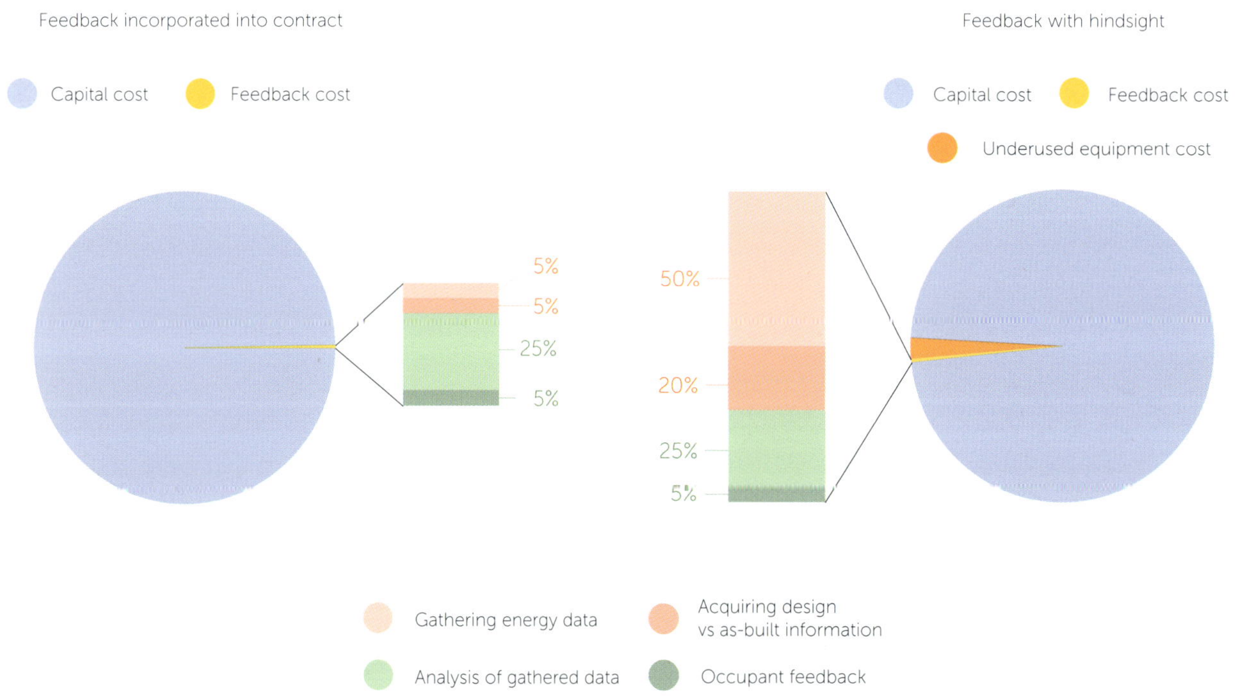

The increased knowledge and enhanced productivity that comes with performance evaluation means that this approach has been voluntarily adopted by forward-looking practitioners. Managing expectations from the start is essential, so that feedback is seen as a form of guarantee to deliver value for the client, the end users and the wider public. Architects pioneering these processes attempt to undertake feedback processes on their projects where they can, because of the knowledge they take forward from one project to the next, and increased confidence that they can deliver buildings that will perform. This increased reputational value 'pays for itself'.[10]

Expertise and skills

Although feedback processes can require some specialist expertise, the majority of tasks are easily within the existing skill set of the architect – see notes 4.01. and 4.02. At the time of writing, architects are the single largest group of professionals leading BPEs. Practices engaged in performance evaluation benefit from hands-on learning, good client relationships and practical research feeding into evidence-based design.

However, either a degree of upskilling is required or the hiring of expertise from a new generation of professionals, who are now becoming well-versed in feedback processes. The basic principles of feedback are slowly being integrated into architectural school curricula, and several universities are developing courses to help practices improve their feedback skills and competencies.[11]

Another approach is collaboration with academic researchers, who can offer an objective view and lend credibility to findings. Several practices engaged academic researchers to work alongside staff to undertake feedback studies.[12] Academic collaboration can lend depth and independence to feedback studies while reducing costs, particularly when the academic partner's research is aligned with a practice's needs. Such symbiotic relationships can be strengthened by publishing findings in a constructive and positive manner through peer-reviewed publications.

A useful approach is to start with light-touch performance evaluation, relying on external support as and when needed. Architects can develop their own feedback methods tailored to their business and the circumstances of a particular project.[13]

For practices experienced in performance evaluations, the next step is to promote their ability to deliver buildings that perform to agreed energy and IEQ indicators and occupant satisfaction parameters. Unsurprisingly, clients increasingly seek out project teams that can demonstrate a track record for building performance. Expert building performance architects can offer their feedback expertise to others to create an additional income stream, while contributing to the necessary step-change needed to realise actual zero carbon buildings.

4.01: EXAMPLE

SPECIALIST EXPERTISE

Architects are well placed to lead the feedback process, because they already manage client and end user communication and the information flow across disciplines. Only when diagnosing issues with complex building systems does specialist expertise become necessary. Nevertheless, feedback must be undertaken systematically and with appropriate knowledge.

THE ABILITY TO DEMONSTRATE AN UNDERSTANDING OF RISKS AND HOW SUCH PROBLEMS CAN BE RESOLVED ENHANCES RATHER THAN DIMINISHES PROFESSIONAL REPUTATION.

THE IMPERATIVE OF THE CLIMATE CRISIS IS TO RADICALLY IMPROVE THE DESIGN AND PERFORMANCE OF BUILDINGS BY 2030, AND THE MECHANISM OF FEEDBACK OFFERS AN IMMEDIATE ROUTE TOWARDS CHANGE.

4.02: EXAMPLE

UPSKILLING FOR FEEDBACK PROCESSES: ARCHITYPE'S JOURNEY

UK practice Architype's commitment to BPE processes for over a decade has enabled the architects to establish simple internal systems to undertake feedback in-house, and incorporate feedback learning from early designs into later projects. Learning from feedback at five schools they designed, the architects were able to deliver a 70% reduction in space heating and hot water use, with high levels of comfort and user satisfaction at Wilkinson Primary School in Wolverhampton, UK, compared to typical CIBSE TM46 benchmarks[14] – see figure 4.4. Lessons learned from feedback at previous schools included the need for improved airtightness and thermal bridging design, reduced south-facing glazing to control summer solar gain, and a hybrid ventilation system that provides natural ventilation in warmer months and benefits from heat recovery in winter.

4.03: EXAMPLE

ADOPTING FEEDBACK PROCESSES

Several organisations that adopted feedback processes in their workflows used the lessons learned to set up practice-wide R&D initiatives and publish the findings. For example, architects at AHR developed a process for project performance contracts (see chapter 10). This widely recognised initiative helped the architects secure a foothold in a new market. Another example is UK property developer British Land, which published numerous projects through the CarbonBuzz disclosure platform, and was able to demonstrate significant reductions in building energy use across its portfolio over a five-year period.

FIGURE 4.4: OPPOSITE

Architype, Wilkinson Primary School, Wolverhampton (2014).

Public disclosure

Things go 'wrong' on all building projects. However, the ability to demonstrate an understanding of risks and how such problems can be resolved enhances rather than diminishes professional reputation. First, feedback results should be shared internally between project team members so that everyone understands the reasons for successes and failures – see note 4.02. Not only does sharing feedback demonstrate the benefits to clients, it also helps make the case for undertaking feedback in new projects.

Next, lessons learned need to be shared externally. Ultimately, the principal beneficiary of public disclosure of feedback results is the wider industry, and in turn the public and the planet. The imperative of the climate crisis is to radically improve the design and performance of buildings by 2030, and the mechanism of feedback offers an immediate route towards change. Public disclosure of feedback results reinforces the credibility of those measures that work and those that do not, discouraging dead-end methods. When findings are published, any problems should be communicated alongside the actions taken to remedy them. Prior to publication of results, client approval must be agreed in writing.

Finally, sharing learning from performance evaluation is an expression of thought leadership. Architects and other industry leaders with a reputation for 'walking the walk' increase brand recognition, and this in turn offers a commercial advantage – see note 4.03.

05

BUILDING CONTEXT AND CONFIGURATION

FIGURE 5.1: OPPOSITE
Frederiksbjerg School, Aarhus, Henning Larsen.

Introduction

This chapter explores different contextual factors that impact on building performance. The overarching goal is making the most of passive design opportunities to minimise remaining energy demands, which are then supplied with energy-efficient systems from renewable sources.

To achieve low-energy buildings, every energy 'end use' in a building must be scrutinised and reduced. However, design assumptions about the impact of passive strategies on energy use often work out differently in reality due to a complex mix of influences such as human, construction and operational factors. Collaborative design processes and detailed expert design inputs early on are essential to tailor design approaches to each context. While clients might be reluctant to pay this upfront cost, feedback has highlighted both the direct and the long-term benefits of early expert input.[1]

This chapter outlines clear design strategies and actions for designers to take charge of the actual performance of their buildings from the start of the design process.

Passive design first

A building's performance is affected by design decisions made early in the design process. Getting the basics right in terms of what, where and how to build can significantly enhance occupant comfort and satisfaction while minimising later reliance on mechanical systems.

Passive design principles use site context (location, orientation, solar angle, shading, wind direction and so on) to generate building form and massing, spatial planning and (passive) heating, ventilation, lighting and cooling strategies, and so on.[2]

The more 'work' that is done by the architecture itself to achieve human comfort and satisfaction with reduced energy use, the less is required by energy-consuming systems that also need to be controlled and maintained over time. Flexible and adaptable design strategies are therefore necessary for a resilient future.[3]

This is reflected by the new feedback hierarchy (see figure 5.2), based on the energy hierarchy, which recognises the need to take into account human factors and address user experience (UX) design to close the performance gap.

The following sections briefly set out key passive design considerations that feedback has shown to significantly influence building performance. This section is not intended to be a comprehensive summary of all passive design strategies.

FIGURE 5.2
Zero energy design hierarchy, prioritising passive design principles with the new feedback hierarchy including human factors and user experience (UX).

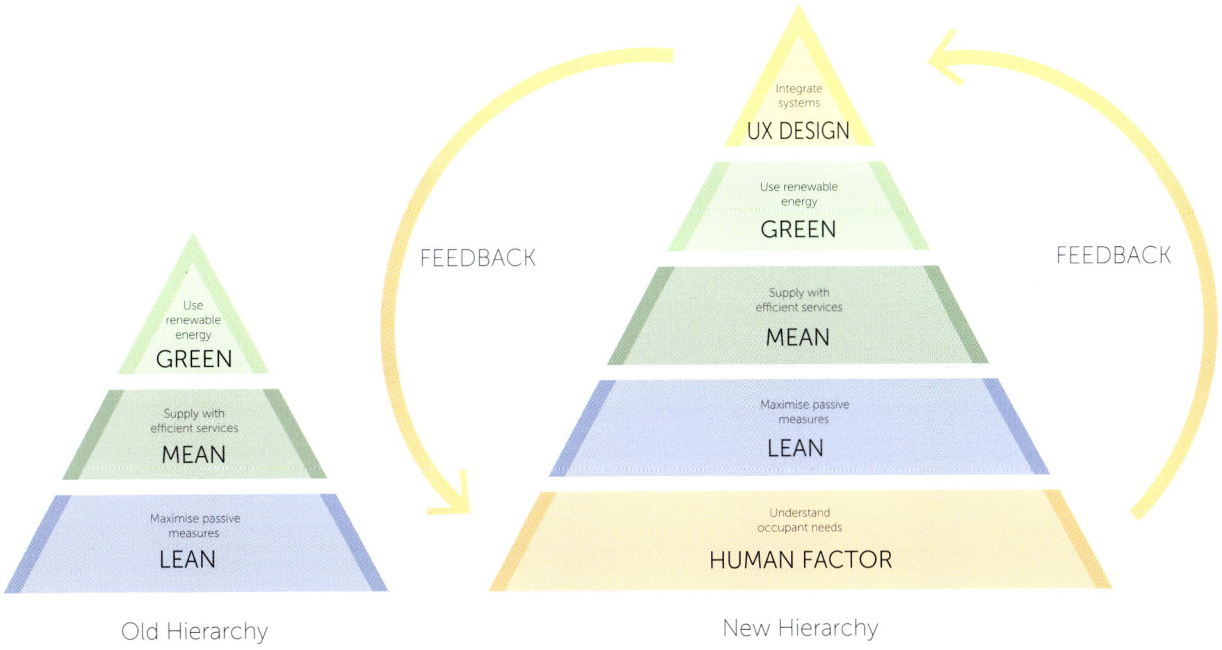

FIGURE 5.3
Urban climate section from the Walkie Talkie building to the Shard, in London.[4] The massing of London's high-rise buildings dramatically impacts their surroundings: the curved facade of the Walkie Talkie (20 Fenchurch Street) reflected sun down to street level, damaging nearby properties and parked cars, until remedial measures were taken. Moreover, the Shard (right) casts a long shadow across its neighbourhood, reducing usability of outdoor spaces and blocking solar access of nearby buildings.

Context and massing

The first step in the design of a new building is positioning it on its site and making the most of its context to minimise energy demand. Good site planning is crucial to optimise and 'lock in' passive design and low-energy strategies at later architectural design stages. Quick contextual microclimate studies lead to a more informed building form which sensitively responds to its context, minimising its impact on the surrounding environment. This includes a careful consideration of urban climate design, because it impacts on a neighbourhood's long-term resilience, as shown by the significant impact of high-rise buildings on their surroundings – see figure 5.3.

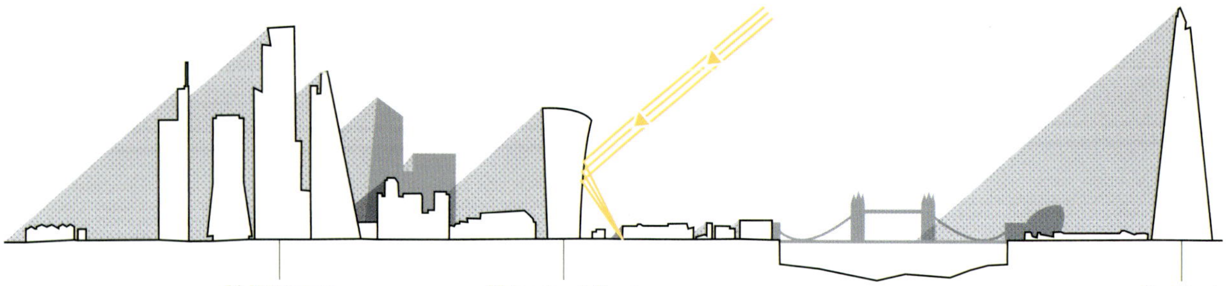

Impact of building form and massing on urban context[6]

Passive measures can also reduce local urban heat island effects through the choice of light-coloured finishes, and, in temperate climates, extensive planting of landscapes and roofscapes. Shaping the urban grid to channel seasonal winds and, for example, positioning water features strategically in hot, dry climates can further improve extreme seasonal conditions – see figure 5.4.

While the local microclimate directly impacts a building's future energy needs, it is only one of many drivers that shape a building's final form (also including adjacent buildings, views, circulation patterns, compactness or form factor, planning constraints, cost and programme, etc.).

Building overheating

Exploiting winter solar gains by optimising orientation, shading, compact form and glazing distribution are widely understood passive design approaches that help reduce the requirement for space heating,[7] but they can also contribute to summertime overheating. With rising temperatures, overheating poses a growing risk to occupant health and wellbeing in commercial, educational and residential buildings.[8] However, overheating is already a problem in some buildings in the current climate.

To illustrate this point, fewer than one in six schools studied as part of the Building Schools for the Future Programme in England had acceptable indoor temperatures, and overheating was highlighted as a particular

FIGURE 5.4 : OPPOSITE
Foster + Partners, Masdar, Abu Dhabi. Climate-responsive urban design including shaded streets, colonnades, planting and water features 'may lower the perceived temperatures on Masdar City streets by 20 degrees Celsius compared to the open desert and by 25 degrees compared to modern Gulf city centres during the hottest days of the year'.[5]

etisa

OVERHEATING RISKS ASSOCIATED WITH COMPACT BUILDING FORMS AND SMALL APARTMENTS ARE COMMONLY OVERLOOKED.

problem.[9] Moreover, overheating risks associated with compact building forms and small apartments are commonly overlooked.[10] For example, overheating is already prevalent in newbuild homes in the UK[11] designed to current low-energy standards.[12]

Because building regulations prioritise tackling heat loss rather than overheating, they inadvertently encourage smaller building volumes, as there is reduced heat loss with less external wall area. Moreover, developers usually want to keep building volume, and especially floor heights, to a minimum to reduce building costs, especially in high-rise residences – see notes 5.01 and 5.02. However, the higher energy use associated with the increased exposed surface area resulting from the taller floor heights can be offset by thicker insulation.[13] And any increase in embodied energy as a result needs to be weighed against improvements in occupant comfort and satisfaction, and reduced reliance on cooling systems.

Taller ceiling heights have also been shown to enhance occupant satisfaction, such as perception of increased desk space and improved thermal comfort[14] – see case study 2 and figure 5.5. Moreover, higher ceilings accommodate ceiling fans to improve summertime thermal comfort and allow windows that bring daylight further into the plan,[15] reducing dependency on artificial lighting, though external shading may be required to block summer solar gains – see note 5.02. These sometimes-contradictory considerations must be taken into account when determining floor heights. Despite tall floor heights contributing significantly to occupant perception of wellbeing, they are often value-engineered out of projects.

FIGURE 5.5

AHMM, Morelands Rooftop, London (2013), section – see also case study 2. The combined effect of higher ceilings, more daylight and better views resulted in measurably greater occupant satisfaction on the upper floor, despite higher temperatures, suggesting that spatial generosity can support occupants' perception of thermal comfort (see 'forgiveness factor', explained in note 2.03).

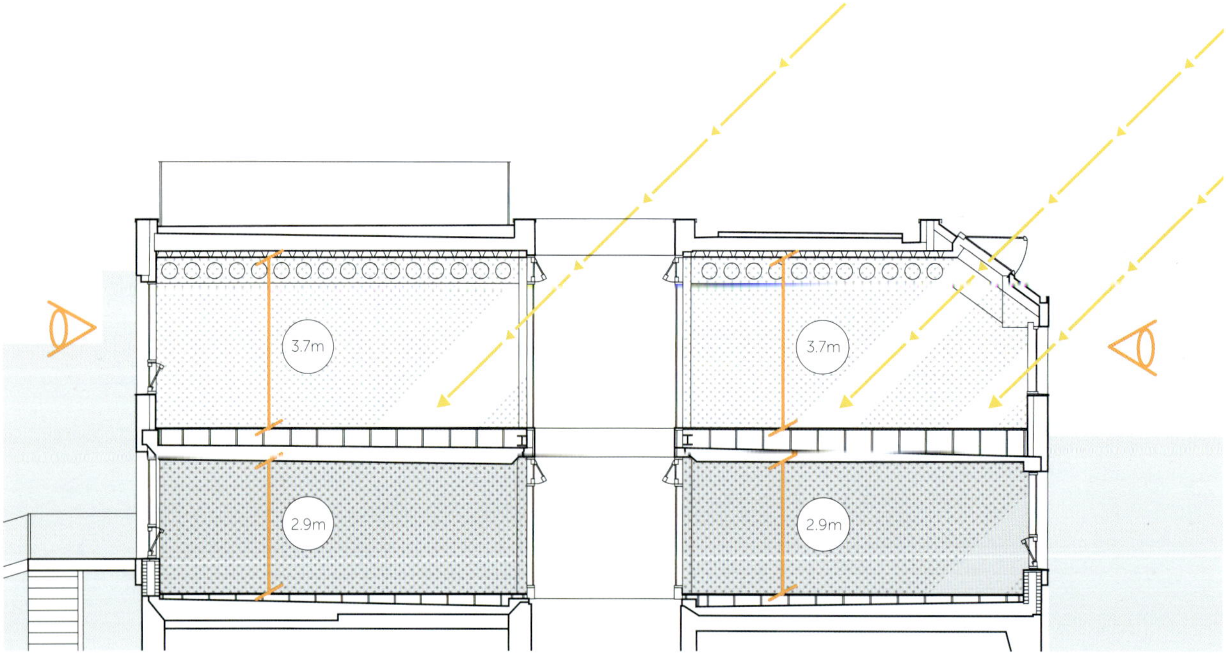

Overheating is not just a human comfort risk, but it can further exacerbate the climate crisis if mechanical cooling systems are retrofitted that are costly to install and operate,[16] leading to a spike in energy consumption in the future. Although few projects currently undertake an overheating risk assessment using future predicted weather data, the noticeable heating of the climate is bringing this type of assessment into the mainstream – see note 5.03. In many northern countries, the use of external shading is relatively new, but it is increasingly necessary. Deciduous trees also have a role to play – see note 5.04. Moreover, some user involvement is crucial in reducing building overheating risk, such as opening or closing windows and louvres.[17] As such, an early design stage understanding of occupants' willingness or ability to help manage overheating risk is important to find the most appropriate solutions for each project, minimising future problems.

Pollution and noise

Addressing outdoor noise or air pollution through passive means is challenging, and in many cases includes a combination of massing, orientation, buffer spaces, planting and careful window placement, among other strategies. However, these passive design solutions are often seen as being costly, not reliable enough, or creating too many design or spatial constraints. However, the replacement by mechanical filtering and supply of fresh air is likely to increase performance risks and associated costs at later stages – see chapters 7 and 8.

An evidence-based approach to massing

To make the most of context as a generator of passive design, these factors must be visualised and quantified early, when it is still possible to adjust and refine a design.

Through early use of environmental simulation tools at concept stage, designers can get quick and useful feedback on the implications of different designs for energy demand – see figure 5.6. However, such environmental analyses and passive design option appraisals are rarely carried out as part of the early design.[18] Instead, such analysis is often undertaken by specialists at the end of concept or detail design stages, when in fact they should be undertaken from the earliest stages of a project (i.e. RIBA Work Stages 0–3).

Some of these analyses can now be performed with mainstream 3D drawing software packages. Until recently, interoperability between 3D modelling and advanced engineering software was a major barrier to carrying out performative analysis of early-stage massing options.

5.01. EXAMPLE

EXISTING BUILDING HERITAGE

Instead of building compact, lightweight new homes, retrofitting historic terraced housing preserves the generous spatial proportions and high thermal mass that are no longer economically feasible to build, and can also result in lower embodied carbon.

5.02: EXAMPLE

IMPORTANCE OF HIGH FLOOR-TO-CEILING HEIGHTS IN PREVENTING OVERHEATING

Feedback studies have shown that more generous floor-to-ceiling heights improve occupant tolerance of summertime overheating, and they increase the resilience of a space to overheating[19] by enabling thermal stratification. When coupled with effective external shading, thermal mass and secure openable windows with suitable operable area, the cooling demands of both domestic and non-domestic buildings can drop substantially.[20]

5.03: EXAMPLE

EVALUATING OVERHEATING RISK IN FUTURE CLIMATE SCENARIOS

Innovate UK's Design for Future Climate Programme has helped establish design standards for assessing overheating under future climate scenarios. Many countries also now provide templates and guidance[21] for future weather data that can be used in thermal analysis software to analyse a building's future overheating potential. Future scenarios should include realistic user behaviour and operating conditions.

5.04: EXAMPLE

IMPORTANCE OF DECIDUOUS TREES

An area that is omitted from compliance calculation and is often neglected by energy modelling is the important effect of deciduous trees. Planting can provide effective shading in summer and let solar rays through in winter while simultaneously providing buffering from pollution and noise, and should form part of a passive shading strategy.

FIGURE 5.6
Digital analysis of direct solar radiation received on a public square and surrounding facades between the months of November and February, to explore opportunities for passive solar heating in massing design. Simulation tools with visual outputs help assess the building performance benefits of different design solutions interactively.

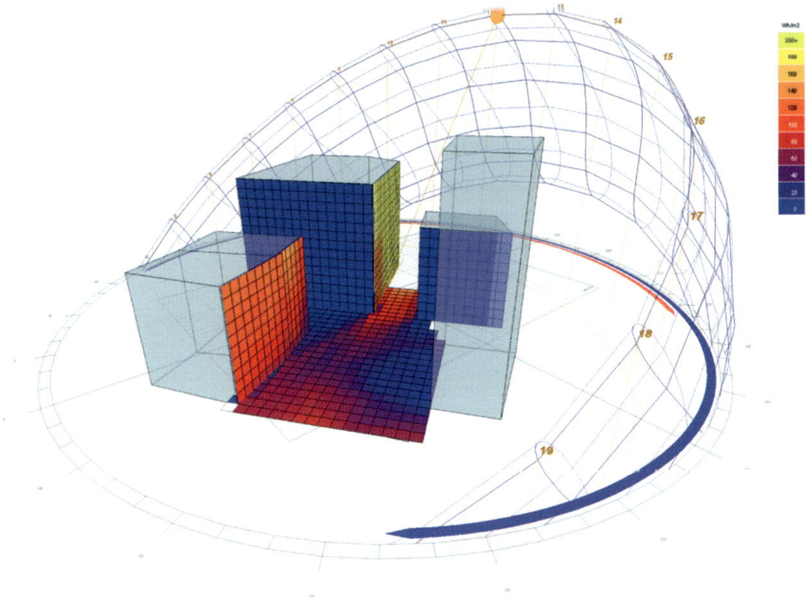

THROUGH EARLY USE OF ENVIRONMENTAL SIMULATION TOOLS AT CONCEPT STAGE, DESIGNERS CAN GET QUICK AND USEFUL FEEDBACK ON THE IMPLICATIONS OF DIFFERENT DESIGNS FOR ENERGY DEMAND

Architectural practices that invest in training to use these methods in-house early on are able to make more informed decisions and retain greater control over different design solutions. This evidence-based approach enables designers to promote passive design proposals to clients with greater confidence.

While it is not necessary to carry out a computer analysis for every project, visualising the relationship between context and building performance validates a designer's options and communicates design intent clearly. For example, solar gain and overshadowing analyses can easily be carried out by hand and inform design development – see figure 5.7.

FIGURE 5.7.
Mikhail Riches, Clay Field, Suffolk, UK. Site section at the design stage: the solar angle generated the architectural form and the space between units, determining roof and wall angles to allow for solar gain and minimise overshadowing.[22]

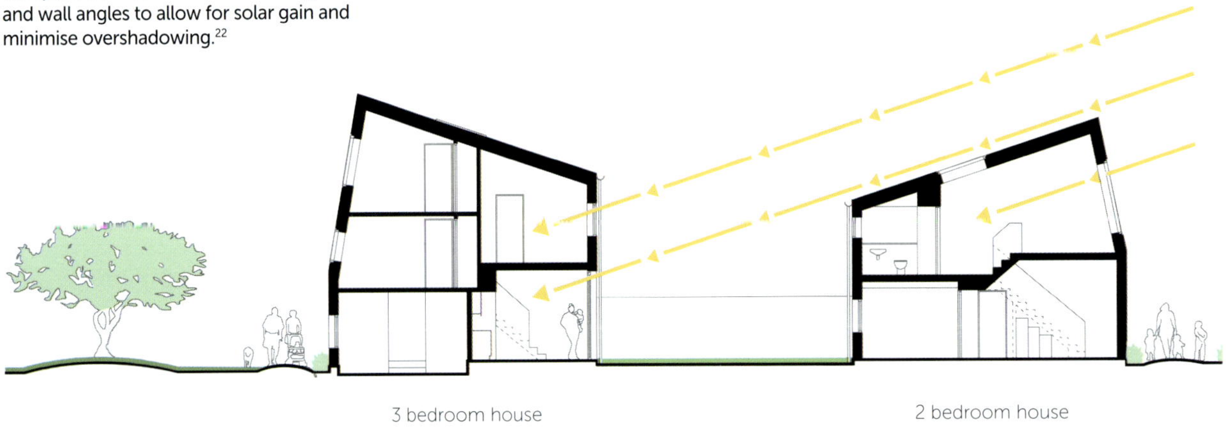

Zoning and adjacency

Zoning is the grouping of spaces or activities together to orient or thermally link spaces and services in a more beneficial way, or to create buffer spaces. 'Adjacency' is the location of spaces in close proximity to one another, for example to achieve effective functional proximity or to enable out-of-hours use for parts of the building. All of these design decisions impact on indoor environmental quality and on energy demand, and are described in this section.

Orientation

Spaces that have low internal heat gains from people, equipment and activities, and that benefit from winter solar gain and changing light conditions, can be located on the south facade if sufficient external summer shading can be provided. This includes offices, classrooms and domestic living areas. In most dwellings, free wintertime heat from solar gain is desirable in living spaces, but not in bedrooms, where overheating is harder to control.[23] In non-domestic buildings, it is customary to have high internal heat-gain functions, such as science labs, trading floors, exercise rooms, catering facilities and conference rooms, facing north, where direct solar gain is minimal. Artists' studios and factories have also traditionally been positioned to benefit from the even lighting of north orientation.

FIGURE 5.8.
AHR, Keynsham Civic Centre, UK (case study 6). Aerial view showing east–west orientation of maximum 15-m deep floors, optimising winter solar gain, cross-ventilation and daylight. The red line indicates surrounding noise levels, mitigated by acoustic louvres in manually operable windows. North–south-facing aspects were maximised through building volume design, minimising low summer sun entering spaces to reduce glare and summertime overheating.

1 Maximum glazing with noise attenuation
2 Sloped roof for PVs
3 Best views captured
4 Narrow footprint enabling cross ventilation
5 East/west solar exposure limited and north/south exposure maximised

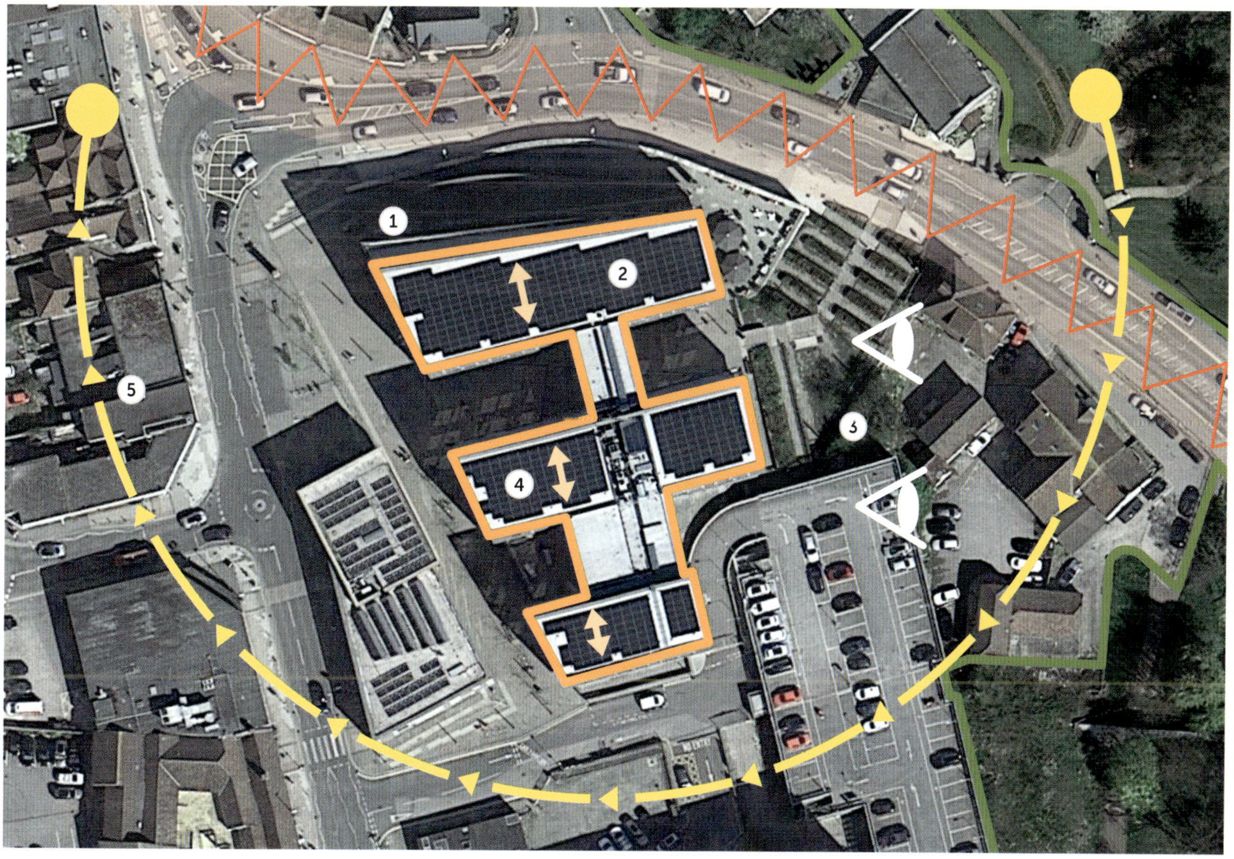

East- and west-facing facades are subject to high solar gains and are difficult to shade from low-angle solar radiation, which can result in glare and overheating. This in turn increases the likelihood of active cooling. For example, in the UK according to one study, north–south-facing secondary schools consume on average 15% less energy than east–west-facing ones.[24] Figure 5.8 highlights how even with an east–west-facing site, north–south-facing aspects can be maximised through building configuration.

Occupant surveys have consistently revealed that glare is a difficult problem to tackle for all orientations – see for example case study 6 . In the northern hemisphere, even north-facing elevations receive low-angle solar rays on late summer afternoons, and low winter sun angles are not always welcome in south-facing spaces. Measures to address this are challenging – see chapter 6 on the design of openings.

Buffer zones

Buffer zones are transitional spaces between one environmental zone and another; often between inside and outside, but sometimes between different internal spaces. Transitional spaces are fundamental to creating a sense of place within a building. While they provide a thermal, acoustic, privacy and spatial buffer, clients often perceive them as unproductive and unnecessary. They are typically entrance lobbies, porches, covered courtyards, atria and corridors, which often form part of natural ventilation routes, too.

Transition spaces can mitigate external weather conditions, leading to reduced energy use, while improving occupants' thermal comfort perception and expectations as they move to their destination in a building.[25] They can also reduce unwanted heat gains and losses, channel daylight, or create an acoustic and olfactory barrier between spaces. In housing, shared stairwells, courtyards and circulation spaces can double as breakout spaces for communal use, as long as environmental comfort can be maintained.

Seamless integration with building services is essential for these spaces to operate as expected. However, budget constraints often mean that buffer zones and circulation areas are subject to value engineering, potentially leading to issues with thermal comfort and daylight of adjacent spaces.[26] For example, feedback studies have frequently found the buffer capacity of transition spaces compromised by underperforming control of heating and lighting conditions.

BUDGET CONSTRAINTS OFTEN MEAN THAT BUFFER ZONES AND CIRCULATION AREAS ARE SUBJECT TO VALUE ENGINEERING, POTENTIALLY LEADING TO ISSUES WITH THERMAL COMFORT AND DAYLIGHT OF ADJACENT SPACES.

Thermal zoning

Designers are often not aware of the important implications of thermal zoning for building services, which is to match spatial layouts to likely occupancy profiles, as well as heating and cooling demands. For example, in dwellings, living areas should be serviced separately from sleeping quarters. This type of clustering is essential in order for buildings to accommodate flexible activities with minimal energy use.

Another example is low-energy office design, where it is customary for naturally ventilated office spaces to be complemented by clusters of meeting rooms or a server room with active cooling provision. Therefore, occupancy patterns, usage schedules and temperature set points of adjacent spaces need to be carefully considered so that these areas can be serviced together to minimise technical equipment, routing, and energy use.[27]

This requires a nuanced understanding of client needs and a building's likely occupancy patterns at design stage, which may not always be possible. Once a building is occupied, inefficiencies can occur if layouts change or energy demand profiles diverge from assumptions because system outputs can adjust to meet the maximum demand within a single zone, increasing energy consumption. For this reason, spatial and services follow-up, flexibility and adaptability are essential.

In complex buildings, different thermal zones sometimes overlap, for example between adjoining atria and corridors, entrance halls and lobbies, or buffer zones. Feedback repeatedly shows that openings which designers assumed would be closed (e.g. café or lobby doors in schools) are left open for access, visual connection or temporary additional ventilation. Therefore, use patterns, opening schedules and temperature set points of adjacent spaces need to be carefully considered so they do not cause systems to fight one another or end up heating or cooling the wrong spaces.

USE PATTERNS, OPENING SCHEDULES AND TEMPERATURE SET POINTS OF ADJACENT SPACES NEED TO BE CAREFULLY CONSIDERED SO THEY DO NOT CAUSE SYSTEMS TO FIGHT ONE ANOTHER OR END UP HEATING OR COOLING THE WRONG SPACES.

Functional proximity

Locating spaces by functional proximity entails thinking about the size, shape and adjacency of spaces in response to occupant needs and use patterns. However, proximity is not always desirable. For example, feedback reveals a mismatch between expectations and outcomes in open-plan living, working and learning, with lack of acoustic separation related to productivity and acoustic comfort.[28] Several of the featured case studies in chapter 9 highlight this issue (for example case studies 2, 4 and 6).

At Frederiksbjerg School in Aarhus, Denmark (case study 5), occupants complained about the lack of focus and acoustic distractions in the open-plan teaching spaces. This was successfully addressed not by introducing acoustic solutions, but by varying pools of pendant lighting to signify quieter learning spaces within the otherwise uniformly lit classrooms, which also reduced energy use.

Out-of-hours use

Feedback from buildings in-use has highlighted that out-of-hours use is a major factor behind unexpectedly high energy use. It is not uncommon for an entire building's technical systems to operate around the clock, when in reality only a small part of a building is occupied[29] outside core hours, using unnecessary energy. This is especially critical for buildings that have a high heat demand or are mechanically ventilated and cooled – see also chapter 8. Therefore out-of-hours use is an important consideration, and activities

FEEDBACK FROM BUILDINGS IN-USE HAS HIGHLIGHTED THAT OUT-OF-HOURS USE IS A MAJOR FACTOR BEHIND UNEXPECTEDLY HIGH ENERGY USE.

that are likely to experience similar occupancy patterns should be located adjacent to one another to reduce the necessary area of out-of-hours use.[30] Successful out-of-hours zoning design should also take access and circulation patterns into account.[31]

For example, education building operators often hire out the auditorium and adjacent lecture theatres, cafes and breakout areas for out-of-hours events. Office buildings increasingly cater for 24-hour occupancy, yet only have a handful of out-of-hours workers present. Successful out-of-hours

ACTIONS	PLAN	DESIGN
	• Highlight the client's responsibility for a building's impact on its wider context and the tangible comfort benefits of improved local microclimate. • Discuss with the client the implications of the zero energy design hierarchy, prioritising passive design principles in the project performance targets (see figure 5.2). • Ensure that overheating analysis, including performance under future climate scenarios, is included in consultants' scopes. • Assess likely occupancy scenarios, including out-of-hours use, ideally with end-user and facilities management (FM) involvement. • Ensure that FM and the design team are adequately trained to undertake feedback processes.	• Include in the environmental impact analysis a building's effect on its local and wider microclimate, highlighting benefits to the public realm. • Carry out several iterations and appraisals of performance analysis of passive design measures, including building mass, orientation, glazing distribution and form factor, to demonstrate the impact of the chosen design on outcomes. • Address risks for overheating and glare through massing and spatial configuration, and optimise floor-to-ceiling heights. • Plan with the client, end user and FM for design proposals to inform spatial clustering, zoning and buffer zones to reflect likely occupancy patterns, and out-of-hours use. • Test overheating conditions for current and future climate, including likely occupant and equipment density. • Design spaces and services to enable future flexibility and adaptability.

zoning design considers access and circulation patterns.[32] The Better Buildings Partnership's Green Lease[33] advocates spatial and system flexibility to implement energy savings, and provides useful tips to incentivise helpful behaviour by out-of-hours occupants.

PROCURE	USE	LEARN
• Incorporate passive design measures influencing the local microclimate, building heat demand and overheating risks in the project Building Performance Register (BPR, see chapter 10). • Ensure that the performance implications of contextual measures and spatial configuration are: • well explained in the BPR • reflected in the drawing set • incorporated into the employers' requirements and contracts • ring-fenced from value engineering • incorporated into building manual and occupant training.	• Collect occupant feedback on comfort and usability conditions of the building context and spatial configuration (including adjacency and zoning). • In the post-occupancy survey, assess implementation of contextual measures and spatial configuration aspects of the design. • Check that thermal zones have been correctly enabled in building control systems, and that they match actual occupancy patterns.	• Consolidate evidence of the impact of a building's architecture on its surroundings and the public realm. • Monitor the impact of spatial configuration on comfort and wellbeing to help quantify the value of spatial quality. • Structure and share feedback to raise industry awareness of the value of the spatial configuration of buildings, and implement lessons learned in own projects.

06

FABRIC FIRST

FIGURE 6.1: OPPOSITE
Everyman Theatre, Liverpool, Haworth Tompkins.

Introduction

The building's envelope dramatically affects both its internal environment and its energy use. Similar in importance to the passive design principles outlined in the previous chapter, the external envelope plays a crucial role in determining a building's overall performance. Getting the building fabric right is absolutely critical to achieving intended performance.

The term 'fabric first' captures the essence of this approach. Fabric first prioritises a building's thermal insulation and airtightness to optimise internal conditions and reduce energy demand, which in turn contributes to reducing the size and complexity of technical systems.

This chapter summarises key design considerations for the building fabric, where feedback has shown they significantly influence building performance and occupant health and wellbeing. It includes different thermal envelope considerations such as airtightness and thermal bridging, as well as feedback on the design of openings for daylight and solar gain while minimising glare.

Regulatory context

Most countries that regulate the thermal performance of buildings set minimum thresholds for whole-building airtightness, and thermal conductivity for walls, floors, roofs, windows and doors, because they have the most significant impact on heating-related energy use, and they also play an important role in occupant comfort.

THERE IS MINIMAL REGULATORY ACCOUNTABILITY FOR THE ACHIEVED PERFORMANCE OF THE BUILDING FABRIC.

While thermal performance is regulated, there is no requirement to validate whether the specified performance has been achieved, nor to demonstrate that it is continuous throughout a building's envelope. In most countries, only airtightness is actually measured and tested as built. This has improved the airtightness of new buildings significantly. However, at present there is minimal regulatory accountability for the achieved performance of the building fabric.

And, while fabric performance is the intense focus of regulatory efforts and design standards,[1] validation from performance in use has not informed these standards. It is generally assumed that what is 'on the box' is what will be delivered.

In addition, it has been found that increased airtightness and tighter building fabric standards can compromise indoor air quality and exacerbate overheating if incorrectly executed or not sufficiently combined with other passive design strategies[2] (see also chapter 5). Other factors, such as vapour permeability, are not even included in compliance calculations or models, so these design solutions are often removed in value engineering. All these parameters must be carefully balanced to create healthy and comfortable environments in buildings that also perform well.

Thermal envelope

The increasing complexity of the building fabric in contemporary construction is a further complicating factor. Most historic buildings feature a load-bearing envelope with high thermal mass, such as stone or brick, which provides weather protection, insulation, vapour permeability and airtightness in a single layer. Today, with few exceptions (see figure 6.2.), newbuild facades are comprised of multiple layers of materials and products, often referred to as a 'build-up'. Adding to thermal envelope complexity are whole-life carbon emission considerations, and the building's end-of-life capacity for dismantling and reuse.

In-situ monitoring of wall U-values and thermal resistances has shown that traditional solid wall construction tends to perform better in reality compared to predictive modelling,[3] and modern, layered construction build-ups tend to perform worse in reality than expected. This is primarily due to inaccurate assumptions about materials and details at the modelling stage, or on-site constructional issues.[4] Feedback is clearly fundamental to understanding the actual performance of the envelope to have a realistic idea of the performance of existing buildings, and how newbuild constructions will perform.

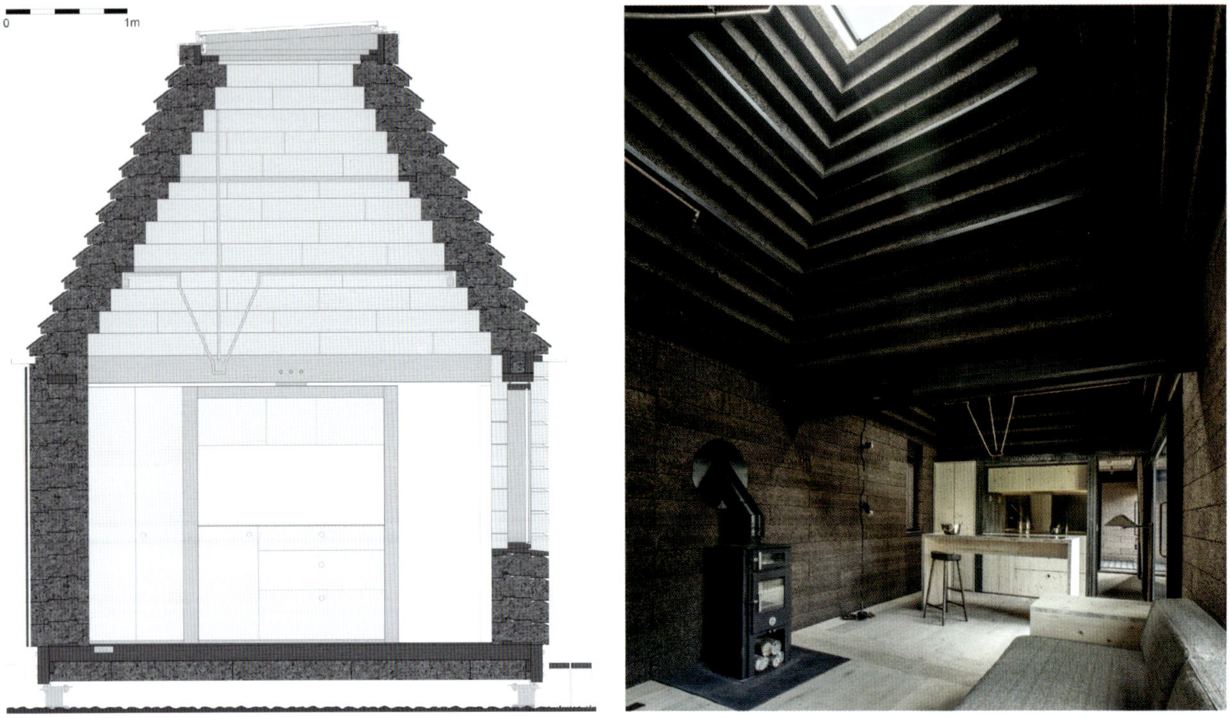

FIGURE 6.2
Matthew Barnett Howland with Dido Milne and Oliver Wilton, Cork House, Berkshire, UK (2019). Conceived as a manifesto for regenerative design and design for disassembly, the Stirling-shortlisted Cork House is built of monolithic cork blocks. This contrasts with the composite building envelopes which dominate contemporary construction and often cannot be disassembled for reuse or recycling.

Airtightness

Airtightness is the first line of action to improve thermal performance, because most heat in buildings is lost via unwanted air leakage due to discontinuities in the building fabric. Feedback from dwellings has shown that an airtightness of 5 $m^3/(m^2.hr)$ at 50 Pa was easily achievable on most projects,[5] and a marginally higher 6 $m^3/(m^2.hr)$ at 50 Pa on non-domestic projects.[6] However, in the domestic sector a third of projects targeting ambitious airtightness levels below 3 $m^3/(m^2.hr)$ at 50 Pa were found to have approximately double their target in reality.[7]

Feedback also showed that airtightness deteriorated over time[8] highlighting that design teams should consider how building performance changes over time and how this might affect heating and ventilation systems.[9] However, some achieved and maintained good airtightness after 12 months – generally those certified to Passivhaus standard, likely due to its framework designed to promote high construction standards – see table 3.1.

Good airtightness can be achieved through application of a continuous air barrier, usually on the warm side of the insulation layer, combined with rigorous detailing and quality construction. As recommended by Passivhaus guidance, drawing a red line on plans, sections and details to represent the airtightness barrier is a clear way to communicate the critical importance of a continuous barrier to all project team members – see figure 6.3.[10]

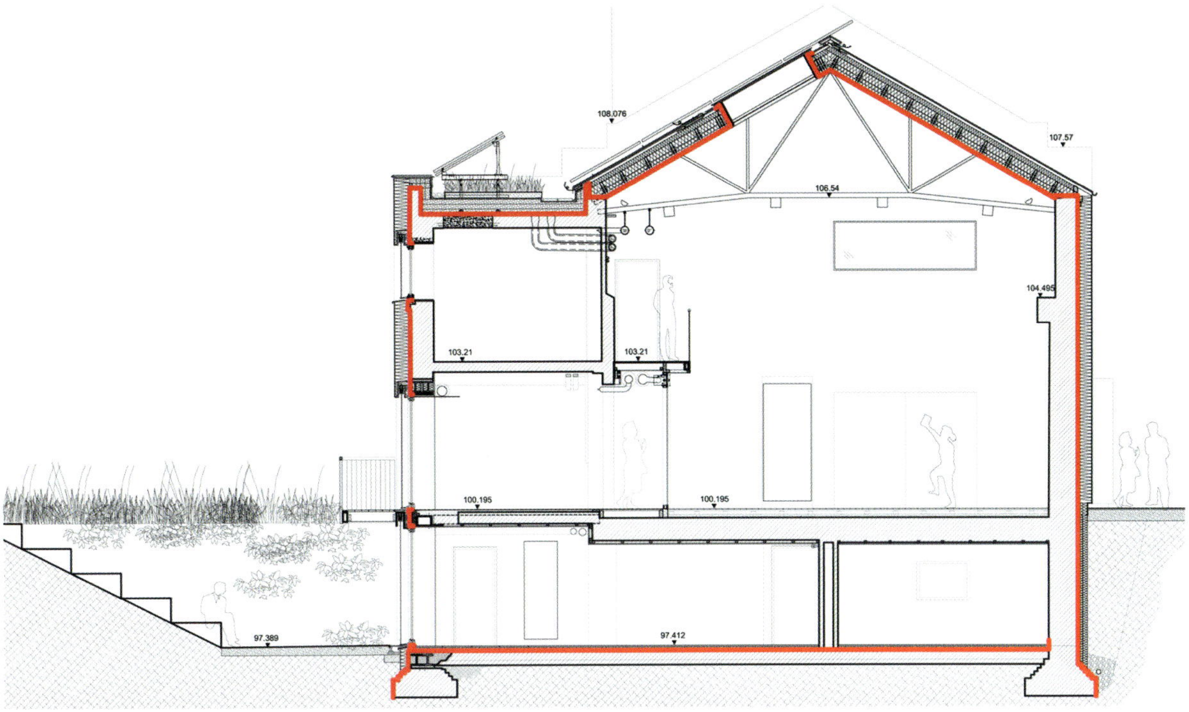

Thermal resistance

Increasing thermal resistance through increased insulation of the building envelope is the most widely recognised way to reduce fabric heat loss – see figure 6.4. Its thickness, material properties, position and fixing all impact performance. Both the position and the materiality of insulation in the fabric build-up are critical to avoid interstitial condensation.

Feedback data shows that glued insulation products risk delamination over time. Mechanically assembled build-ups lend themselves better to disassembly, reuse and recycling; however, thermal bridging risks need to be managed.

FIGURE 6.3

bere:architects, Mayville Community Centre, London (2011). Section with red line indicating continuous air barrier, which should also be consistently drawn on all construction details to communicate to the rest of the design team and contractors.[11]

FIGURE 6.4

bere:architects, Passfield Drive, London, UK. Thermal image of a Passivhaus terraced house showing the impact of fabric insulation compared with adjacent houses.

Vapour permeability

Humidity levels vary between indoors and outdoors, and a vapour-permeable or 'breathable' building envelope can passively buffer some of this difference across the building fabric (see case study 3). Generally, unwanted moisture in buildings is a key source of occupant health issues and fabric failure, so managing moisture in the building fabric is essential to extend the life span of buildings. For this reason, the choice of a sealed or a vapour-permeable envelope is an early strategic decision that affects material and building choices.

Some building materials can help to stabilise indoor air humidity through their ability to temporarily absorb water vapour.[12] Feedback analysis has also shown that the ability of hygroscopic materials to absorb and release moisture can significantly change their thermal conductivity, as well as their capacity to store heat.[13] However, vapour-impermeable materials such as plastic foam insulation products are often specified, because they are thinner, lighter and quicker to install than mineral or fibre-based products with similar thermal resistance.

The hygroscopicity of the building fabric is often poorly understood and communicated by designers, with the result that the building fabric can be compromised by value engineering or poor workmanship.[14] Moreover, vapour-permeable design strategies are too often overlooked, as the vapour permeability of materials is not part of compliance calculations and is rarely included in dynamic thermal modelling.

6.01: DEFINITION

THERMAL BRIDGE

A thermal bridge is an area or component of the external fabric that has higher thermal conductivity than the surrounding materials, creating a path of least resistance for heat transfer. Thermal bridges typically occur at junctions between build-ups and around penetrations. Areas prone to thermal bridges include windows, doors, balconies, lift shafts, service risers and fabric penetrations.

Thermal bridges

Similar to the airtightness barrier, continuity of the insulation layer is critical to avoid thermal bridging. Thermal bridges can cause up to 25% of a building's heat loss, yet feedback shows that this detail is often not followed through rigorously enough during predictive and/or compliance modelling.[15]

Thermal bridges are eliminated by continuous insulation (see figure 6.5.) and by detailing a thermal break into the envelope at the design stage. In detailing thermal breaks, designers must anticipate whether the complexity of construction matches the skills likely to be available on site. The location of service ducts, pipework and wiring must be anticipated and clearly communicated to avoid on-site changes, because ad-hoc service penetrations are prone to leaks. Late modifications often necessitate additional bends in pipework, which can increase heat loss and additional energy use. Passivhaus construction standards are rigorous in achieving intended airtightness, thermal resistance and the elimination of thermal bridging.

THERMAL BRIDGES CAN CAUSE UP TO 25% OF A BUILDING'S HEAT LOSS

Thermal mass

Intelligent use of exposed thermal mass can help keep buildings passively cool in summer and balance winter heating needs, supporting the thermal comfort of occupants by making use of the thermal capacity of

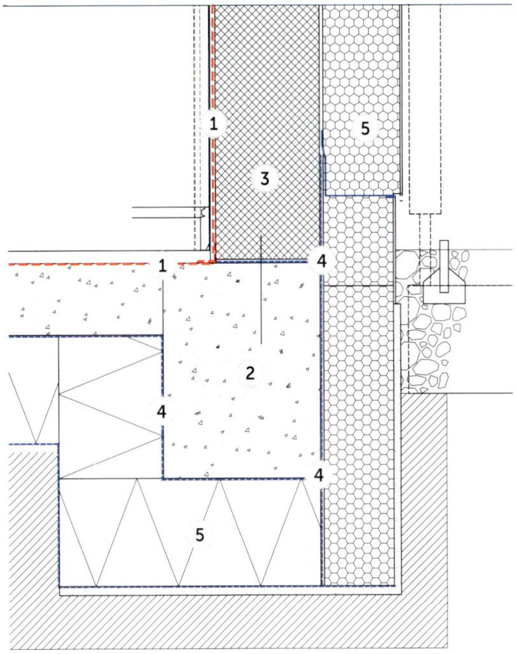

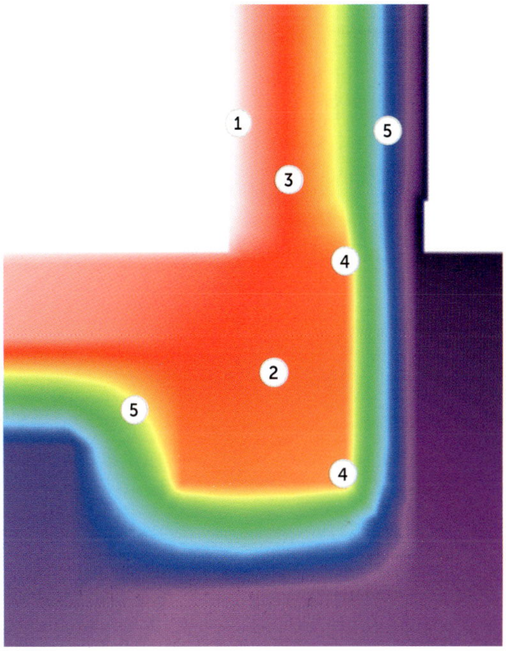

FIGURE 6.5.
Hamson Barron Smith, Carrowbreck Meadow, Norfolk, UK (case study 3). Foundation detail and thermal bridge study, highlighting how continuous insulation layers can minimise thermal bridging.

1 Airtight layer
2 Concrete slab
3 Clay block
4 Vapour barrier
5 Insulation

dense materials (case studies 1, 2 and 4). The use of thermal mass is not recommended where it risks retaining too much heat in summer that cannot be dissipated at night. Without secure and effective night-time cooling, thermal mass can contribute to overheating instead of reducing it.

Feedback from building performance evaluations has revealed frequent misunderstanding of the role of thermal mass. For instance, thermally massive materials have often been found covered up with gypsum board, compromising their beneficial effect.[16]

Although thermal mass has a major impact on building performance,[17] it is not robustly accounted for in building regulations. This may be due to the fact that most compliance models cannot reliably predict its benefits. For example, calculating the impact of thermal mass requires dynamic analysis, because it relies on evening out temperature changes over a 24-hour cycle. Yet regulations only require steady-state models (see chapter 4). Steady-state models also largely overlook the seasonal effects of thermal mass,[18] and even dynamic models have been shown to poorly estimate its buffering capabilities.[19]

Troubleshooting poor fabric performance

From missing insulation to poor positioning and fixing, construction quality issues in the building fabric are widespread and are a major contributor to the performance gap. For example, a study of 76 new homes in the UK found that carbon emissions were on average two to three times higher than initially estimated.[20] This discrepancy was not solely due to fabric issues, and included teething problems with new technologies and how to

A STUDY OF 76 NEW HOMES IN THE UK FOUND THAT CARBON EMISSIONS WERE ON AVERAGE TWO TO THREE TIMES HIGHER THAN INITIALLY ESTIMATED

6.02: EXAMPLE

WINDOW DESIGN CONSIDERATIONS

Windows form a visual and physical link to the outside world, and affect just about every aspect of a building's performance, be it thermal and visual comfort, air quality or acoustics. This is why, when designing window openings, numerous issues must be considered: views and passive supervision of outdoor spaces; lighting conditions (solar gain and glare); fresh air provision; air movement; thermal, auditory and olfactory conditions; and ingress/egress of pollution.

use them.[21] However, incomplete insulation, gaps in air barriers and thermal bridging were common, while in one case, the insulation was entirely missing.[22] Similar issues were found across 50 UK non-domestic projects, where installed insulation was often reduced, missing or of different composition than what had been approved by building control.[23] The use of thermal cameras at the construction stage could have highlighted these problem areas in time for them to be easily rectified – see figures 6.4. and 6.8.

To ensure good performance, all the factors that affect fabric performance must be identified upfront and tracked across the life of a project. Additionally, the entire project team must be made aware of the critical impact of construction details and specifications on performance targets (see case study 3), and the same must be communicated clearly to subcontractors.

Openings – windows, doors, rooflights and services

Building openings present the greatest design challenge to building performance outcomes, as every opening presents a performance risk, be it a window, door, rooflight, duct vent or pipe penetration. Openings include prominent architectural features that dramatically influence the way people experience buildings. See note 6.02.

Windows and openings can be among the highest-value construction packages on a project, and feedback shows that they significantly impact a building's resource consumption, indoor environmental quality and the satisfaction of occupants.[24] But surprisingly, window performance is one of the least-studied elements in most building performance evaluations (BPEs). If window design is to perform to expectations,

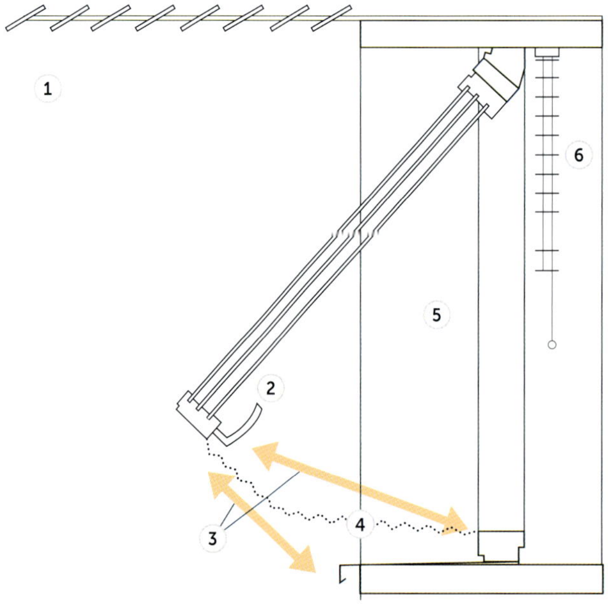

FIGURE 6.6

Diagram of a window detail showing the complexity of an operable window with louvres, blinds, a deep reveal, ironmongery and an insect mesh all affecting performance. See also chapter 5.

1 Louvres

2 Handles

3 Calculated and actual operable area

4 Insect mesh

5 Window reveal

6 Blinds

architects must consider how people actually use windows, and match design assumptions to the realities of human experience – see figure 6.6. Feedback reveals numerous performance shortfalls around window frames and joints, such as reduced thermal resistance and airtightness.[25]

Daylighting, solar gain and glare

The balance between daylighting, desirable solar gain and undesirable glare is difficult to get right. Orientation plays an important role; however, some form of external shading is beneficial to most buildings to ensure good daylighting without glare and prevent overheating.

External louvres are often considered too inflexible to accommodate changing conditions, unless they are automated, which makes them costly. The technology to animate external components has developed rapidly, with buildings such as AHR's Al Bahr Towers in Abu Dhabi demonstrating state-of-the art motorised louvre design – see figure 6.7. In addition to high capital cost, commissioning and maintenance costs are a key consideration in automation – see chapters 7 and 8.

Where window design and external louvres do not completely eliminate glare, internal and/or inter-pane blinds can be specified to improve visual comfort. If responsibility for controlling blinds is not clear, feedback has shown that they are likely to be inadvertently left closed for long periods, with repercussions for electric lighting (see case study 2 and note 6.03).[26] To ensure that blinds are used correctly, proactive communication between occupants and FM is needed – see note 6.03. Feedback also found that both manual and automated blinds are a frequent source of occupant discomfort and frustration, as well as maintenance problems.[27]

ARCHITECTS MUST CONSIDER HOW PEOPLE ACTUALLY USE WINDOWS, AND MATCH DESIGN ASSUMPTIONS TO THE REALITIES OF HUMAN EXPERIENCE

6.03: EXAMPLE

USER OPERATION OF BLINDS

While at Morelands Rooftop office (case study 2) feedback reported no issues with external blinds, at Keynsham Civic Centre (case study 6) occupants frequently left internal blinds partially or completely closed, which triggered unnecessary use of artificial lighting. At Carrowbreck Meadows (case study 3), when residents were asked how easy it was to use their external blinds, 85% responded 'very easy' and the remainder 'quite easy.' Careful design consideration and in-use communication and follow-up are required. See chapter 9.

FIGURE 6.7

AHR, Al Bahr Towers, Abu Dhabi (2012). Al Bahr Towers' motorised folding louvres provide translucent shading to all but the northern glazing. As the sun moves around the buildings, the louvres gradually open and shut, based on the intensity of the solar radiation hitting the facade.

Fresh air

During warm periods, window openings are the main source of fresh air and air movement in naturally ventilated buildings. Operable windows are commonly sized to provide the minimum air-change rates required by legislation. However operable area should also be sized to allow for an adaptive comfort strategy (see chapter 2), which requires higher air-change volumes.

Window size and position in the facade are critical to enable good air flow, but more subtle parameters such as window type and depth of reveal also impact performance – see figure 6.6. For example, while top-hung windows are rainproof, they do not provide the same volume of airflow as a side-hung or a sash window, even with the same operable area. Feedback has also shown that the depth of window reveals, critical for solar shading, also significantly impacts air changes, but this is rarely accounted for in thermal models. This resulted in reduced ventilation area in reality.[28] Similarly, safety restrictors, acoustic grilles or insect mesh on windows restrict air flow – a factor rarely incorporated into thermal models, which can reduce air flow and exacerbate summertime overheating and poor air quality[29] – see figure 6.6 and 6.8.

When windows are manually operated, ease of use is critical. Feedback revealed frequent examples where handles were out of reach, of poor

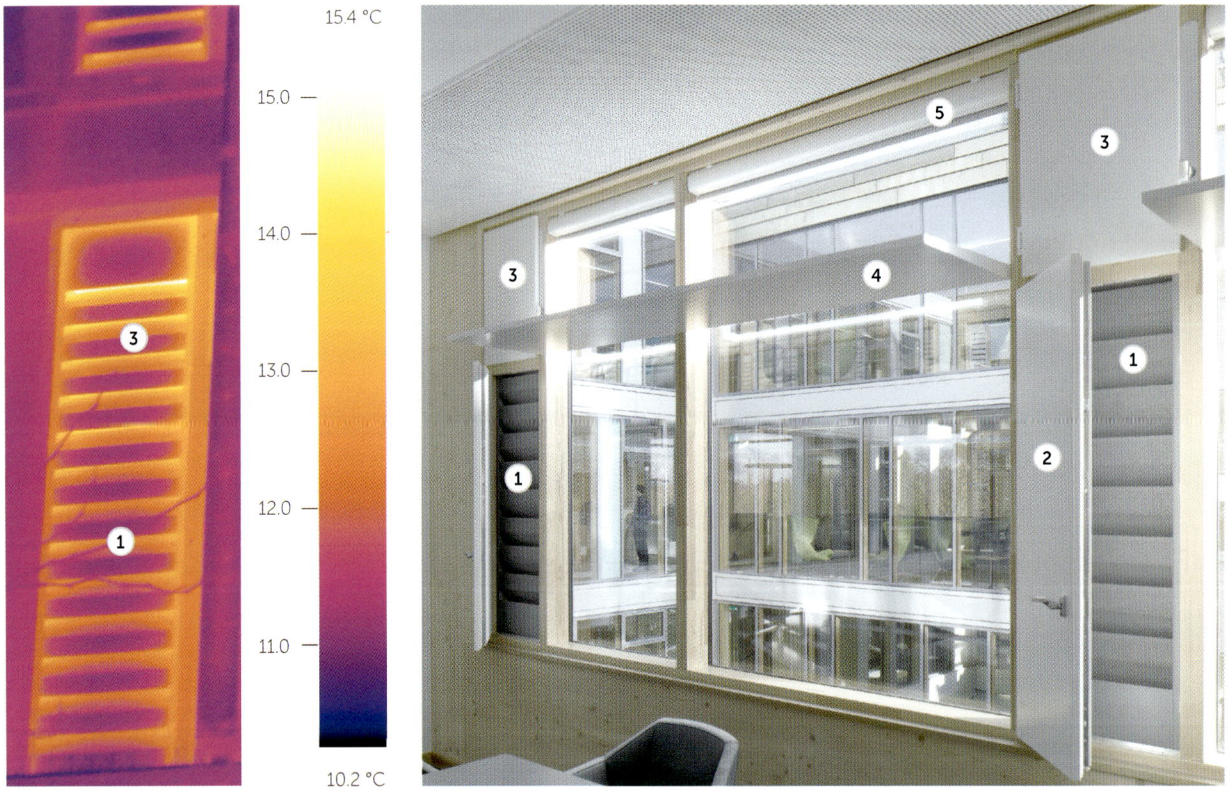

FIGURE 6.8

Keynsham Civic Centre, UK (case study 6). Naturally ventilated buildings often require more complex window design to enable fresh air without compromising security, acoustics or indoor air quality. The interior image shows the window and louvres, while the infrared image on the right shows thermal bridging at the acoustic louvres.

1 Acoustic louvres

2 Manual openings

3 Automated ventilation panes

4 Light shelves

5 Blinds

quality or improperly adjusted or the opening mechanism not properly weighted. When windows cannot be easily operated, they are rarely opened or left open excessively – see chapter 8. This can cause poor indoor air quality, overheating or unnecessary heat loss.[30] Where openings are automated, actuators present a major risk of failure or misuse, prone to issues such as being jammed, too noisy, too high-maintenance or too stiff, or offering a lack of control over how much they open, or having switch problems.

Solar gain and light transmission

The defining properties of a glazing system are its light transmittance (g-value), visual transmittance and thermal transmittance (U-value). The latter is specified either as centre pane or an average for the unit. To achieve optimal performance, it is important to consider the entire unit (i.e. glass and frame) when specifying windows.

Until recently, it was considered best practice to specify the highest visual transmittance with the lowest g-value for commercial buildings, and the highest visual and solar transmittance in housing. However, with increased overheating risk, architects now look to balance winter heat loss and summer solar heat gains, specifying windows with both low U-values and low g-values. However, low g-value glass could be noticeably tinted, which might lead to darker spaces and unintentionally increase use of artificial lighting – see note 6.04.

Triple glazing

The thermal conductivity of glazed openings can be greatly improved by triple glazing in well-insulated frames. Although currently considered costly, triple glazing offers the additional advantage of improved insulation from external noise such as traffic.

The hanging of heavier glazed openable windows and doors requires extra support and adjustment after the first 9–12 months of use to ensure a good seal. Without this fine-tuning, frames can sag and damage the seals. Feedback has shown that even when rehanging is specified by architects or manufacturers, it is frequently value-engineered out or overlooked, leading to underperforming expensive building components, with usability and airtightness compromised.[31]

Acoustic conditions

Outdoor noise, particularly traffic on busy roads, is a common driver in opting for mechanical ventilation. However, the use of acoustic louvres in operable window panels can successfully attenuate noise and enable natural ventilation even in noisy settings – see figure 6.8. and case study 6. Actual performance data should be requested from manufacturers to evidence airtightness and thermal resistance and to include the actual air-change rates of the attenuated openings in building energy modelling at design stage.

6.04: EXAMPLE

LIGHT TRANSMITTANCE

At the UEA Enterprise Centre (case study 7), low g-values were specified to control solar glare. In rooms with task lighting and no overhead lighting, feedback showed that certain north-facing rooms with trees outside were too dim, particularly in winter.

ACTIONS	PLAN	DESIGN
	• Incorporate clear targets for high-performance thermal envelope and openings in consultants' scopes. • Incorporate into client's brief the requirement to define and track how these targets will be achieved. • Consider a vapour-permeable fabric and the use of thermal mass if beneficial. • Consider external shading unless shown to not be necessary – involve specialist consultants and subcontractors early. • Clarify with the client unambiguous lines of responsibility for as-built fabric performance, requirements for additional communication, and quality control protocols, such as a Building Performance Register (BPR) – see chapter 10. • Incorporate post-completion thermal imaging and surveys into relevant consultant scopes.	• Follow Passivhaus guidance for thermal resistance and airtight building design and construction, including detailing of perforations and thermal bridging. • Analyse fabric performance and window openings using both summer and winter weather data and under future climate change scenarios to balance daylighting, heat loss and overheating risks. • Take into account realistic occupancy use scenarios and actual performance of openings, reveals, materials and so on when undertaking building energy modelling. • Ensure that indoor environmental quality (IEQ) targets can be delivered by window design, and consider how louvres, blinds and openings will be controlled, and by whom. • Highlight the location of thermal breaks, perforations, and wind, air and vapour barriers clearly for all drawings and details. • Plan ahead for in-situ monitoring, and specify the implementation of appropriate sensors. • Carry out a risk assessment to include value engineering, availability of specialist subcontractors and maintenance, commissioning, and user-friendliness of interfaces.

PROCURE

- Ensure design measures affecting thermal performance are communicated in contracts and documented to enable effortless management.
- Ensure that the main contractor takes responsibility for the performance and defects relating to any openings, and that they are checked and adjusted after one year as part of the supplier's contract.
- Carry out thermal imaging as soon as the envelope is sealed but before finishes are applied to identify any discontinuities in thermal, air and vapour barriers, and remedy any ill-fitting seals.
- Inspect on site and document that build-ups are as prescribed and that insulation, airtightness and vapour barrier layers join up across junctions.
- Assess any on-site changes in design and specification, and assess how these changes affect performance.
- Include in the building manual the requirement for any training and maintenance, as well as as-built drawings, specifications and user instructions.

USE

- Carry out thermal imaging to identify any discontinuity in the thermal, air and vapour barriers – to include all openings and junctions.
- Carry out possible further tests for diagnostics.
- Check implementation and working of user-friendly interfaces to enable effortless management of openings and penetrations.
- Train end users and maintenance staff in the use and maintenance of openings and penetrations.
- Monitor seasonally the IAQ and the operation of windows and doors once the building is in use, and identify and rectify issues.
- Include relative humidity levels in monitoring if possible, and check for signs of condensation around junctions.
- Include seasonal validation of indoor environmental quality (IEQ) by monitoring of indoor conditions and occupant satisfaction surveys.

LEARN

- Document factors contributing to defects and successes in construction quality.
- Document how the validation of thermal performance via thermal imaging identifies defects early on, and how these are rectified.
- Document how performance contracts and processes need to be improved to achieve better build quality, and implement these in future projects.
- Share the lessons learned and the results of feedback in the public domain.

07

INTEGRATING TECHNICAL SYSTEMS

FIGURE 7.1: OPPOSITE
Storey's Field Centre, Cambridge, MUMA.

Introduction

Technical systems use energy to top up a building's passive supply of fresh air, heat, coolth and daylight. Specifying the right systems and ensuring they operate correctly is critical, because underperformance can jeopardise zero energy targets, amplify issues with indoor environmental quality (IEQ) and increase operating costs.

Whether or not a building needs cooling is an important determinant when deciding to opt for mechanical ventilation. The decision between natural, mixed-mode or mechanical ventilation impacts enormously on the spatial, services and envelope design of buildings.

Paradoxically, the recent drive for energy-efficient buildings has accelerated mechanisation: for instance, smart controls can vary the supply of fresh air, heating and cooling according to demand, and make it easier to achieve regulatory compliance than natural ventilation does. Natural ventilation, on the other hand, requires more complex calculations, greater restrictions on building configuration and a coordinated approach between design disciplines, while offering less certainty in avoiding overheating.

This chapter highlights the common pitfalls with energy-consuming heating, ventilation and air conditioning (HVAC) technical systems that have emerged from feedback studies, and shows how these can be successfully addressed.

Ventilation and air-conditioning

Since the 1970s, energy consumption in buildings has doubled, and in many countries almost half of all energy used in homes is for heating and cooling. Additionally, growing prosperity in hotter climates as well as a changing global climate have accelerated a trend for mechanical cooling – and where there is already air-conditioning, operating times might increase.[1]

Mechanically ventilated and sealed buildings often consume more energy than those with naturally ventilated or hybrid solutions (see note 7.01), but they are favoured for their ability to support large variations in occupant density and behaviour, hours of use and equipment density. In contrast to this, passive design solutions require more careful design coordination to provide occupant comfort, with features such as operable windows, higher ceilings, increased insulation, shading and noise attenuation – see figure 7.2 and chapters 5 and 6.

In large non-domestic buildings, thermal comfort can be provided together with indoor air quality (IAQ) by a centralised HVAC system, often across multiple thermal zones, allowing cellular and open-plan areas to be positioned anywhere on a floor plate – see figure 7.2. However, this is expensive, energy and maintenance-hungry, and can present certain risks to occupant wellbeing, related to compromised IEQ such as lack of acoustic, thermal, visual and respiratory comfort (see chapter 2).

7.01: EXAMPLE

COST AND PERFORMANCE OF MECHANICALLY VENTILATED BUILDINGS

Feedback studies show that mechanically ventilated offices and education buildings in the UK typically consume double the electricity of naturally ventilated ones,[2] and the gap between the expected and achieved performance is approximately 40% higher for these buildings – see figure 7.4. International organisations such as the British Council for Offices (BCO) and the Better Buildings Partnership are developing new guidance and standards to reflect changing user needs, placing greater focus on health and wellbeing and reduced resource consumption.

FIGURE 7.2

Diagram showing the spatial configuration of a fully mechanically conditioned, or 'sealed', building with plant rooms and risers (left), and a naturally ventilated building with opening windows and shallower floor plates, higher ceilings, no risers and a variety of optional low-energy and renewable systems (right).

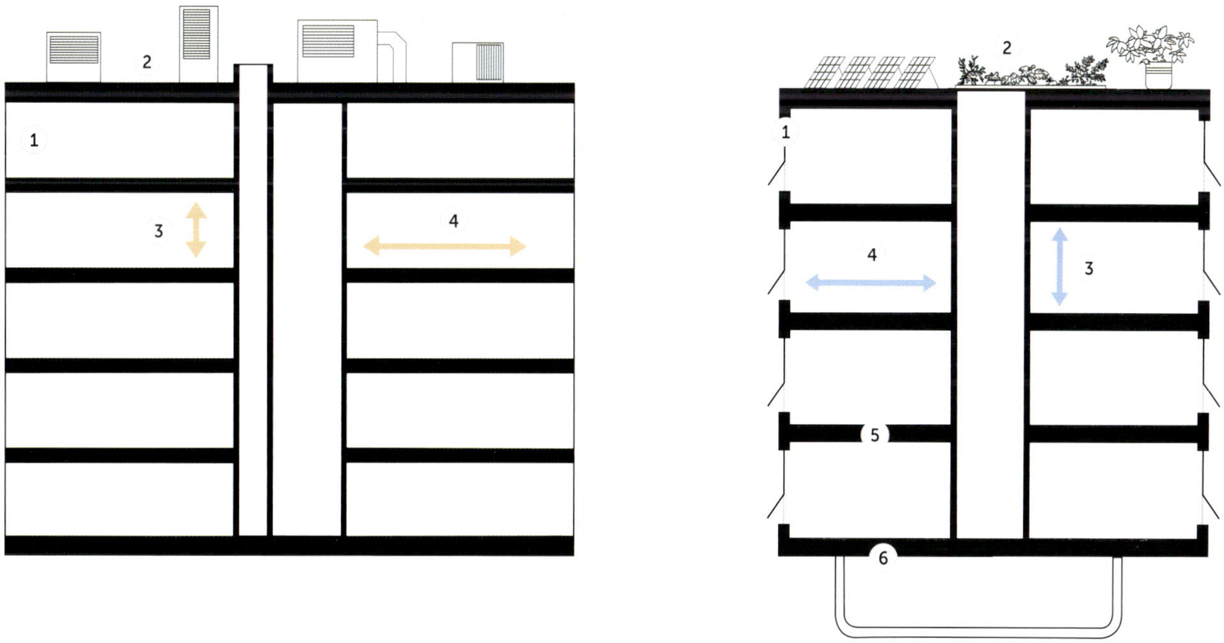

1 Sealed facade
2 Plant rooms and risers on roof
3 Lower ceilings
4 Deeper floor plans

1 Operable facade
2 Room for solar panels and greenery on roof
3 Higher ceilings
4 Shallower floor plans

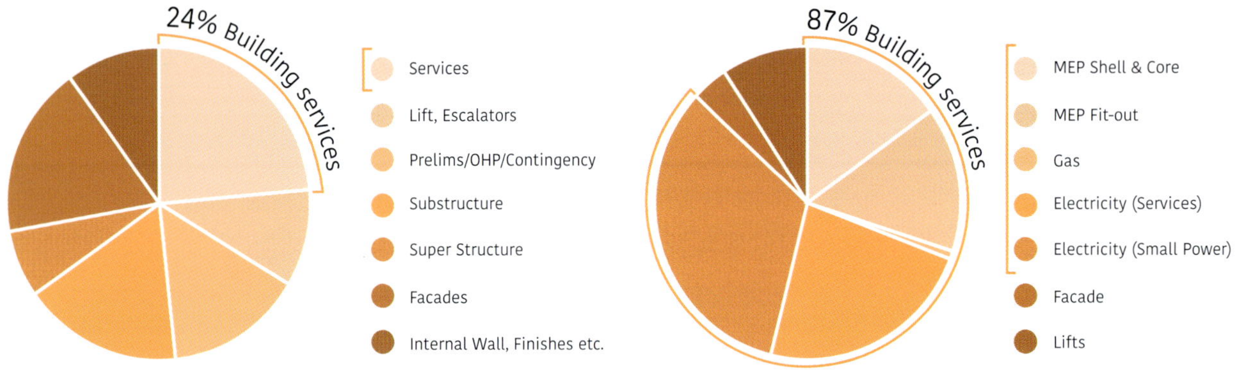

FIGURE 7.3
Capital cost vs whole-life cost of M&E systems in typical commercial offices – as proportion of total cost.[3]

Providing tempered fresh air can be the single highest source of capital and whole-life costs for many building types. In a typical commercial office building, M&E systems account for nearly a quarter of capital cost and over 85% of whole-life cost[5] – see figure 7.3.

Occupants of naturally ventilated buildings tended to be slightly more satisfied with indoor environmental comfort conditions, even if temperature and CO_2 concentrations varied more, and were in some cases higher than in buildings with full HVAC systems.[6] Anecdotal evidence suggests that occupants appreciate the taller ceiling heights and better daylighting[7] that are common in naturally ventilated buildings – see for example case study 2. Greater personal control over indoor conditions may also play a role.

However, as urban air and noise pollution levels drop due to increased use of electric transport, a passive approach to ventilation is becoming more desirable again. Even in dense urban areas, where noise and pollution levels are high, mixed-mode ventilation, where fresh air flow is attenuated and filtered, is viable for a significant part of the year in temperate climates and can form part of a building's visual identity – see figure 7.5.

PROVIDING TEMPERED FRESH AIR CAN BE THE SINGLE HIGHEST SOURCE OF CAPITAL AND WHOLE-LIFE COSTS FOR MANY BUILDING TYPES.

FIGURE 7.4
Diagram highlighting larger energy performance gap in mechanically ventilated buildings (left) vs naturally ventilated buildings (right).[4]

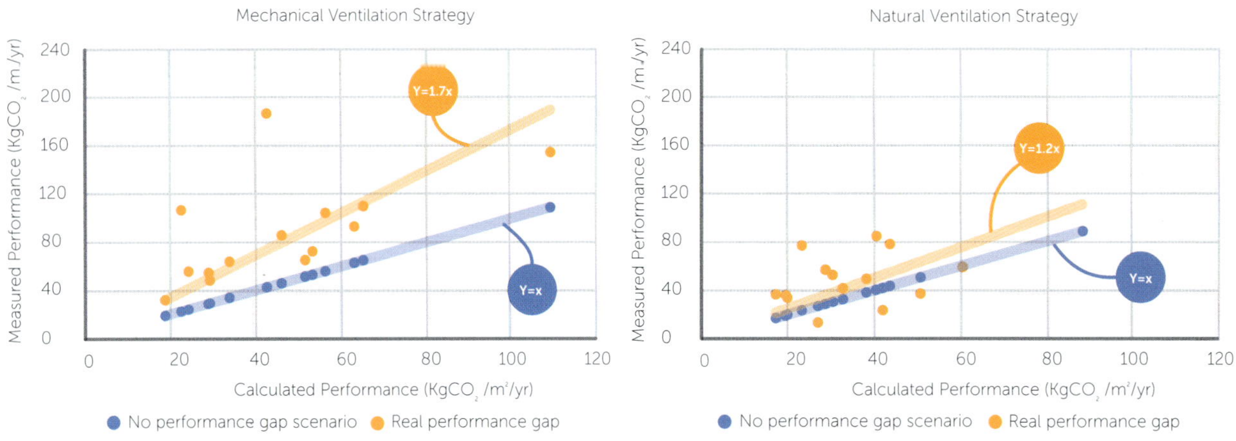

FIGURE 7.5
Foster + Partners, Bloomberg, City of London, 2017. The facade of the Bloomberg building incorporates 117 operable and acoustically attenuated blades that open to naturally ventilate the building when environmental conditions permit. They are supplemented by mechanical systems based on CO_2 monitoring.

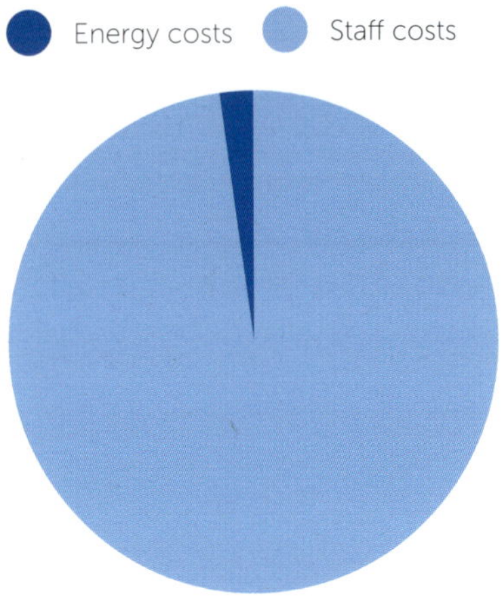

FIGURE 7.6
Staff costs vs energy costs.

Risk of poor air quality

Sealed buildings have been shown to expose occupants to potentially poor IEQ,[8] and feedback processes are crucial in order to understand, fix and avoid these problems. For instance, in one fully sealed school building, a failure occurred in one zone's air-handling units (AHUs), leaving an entire stack of classrooms without fresh air.[9] With no CO_2 monitors, the occupants did not realise the drop in air quality until BPE monitoring highlighted the issue.[10] Despite a requirement for non-domestic buildings to have their ventilation and air-conditioning systems inspected at regular intervals, it is common to find filters unchanged and winter recirculation mode not switched to full fresh air in summer mode.

Air quality is also a major consideration in housing, where better airtightness necessitates a greater reliance on year-round background mechanical ventilation. Even airtight buildings require at least some manual ventilation openings, and a user-friendly alert and filter changing system, as well as occupant training to maintain safe and healthy air quality.

Finally, the COVID-19 crisis highlighted that recirculated air can have potential health effects in the case of airborne transmission of diseases.[11] In such cases, the best advice at the time of writing is that ventilation systems should be operated by increasing the rate of air change, increasing the use of outdoor air, and decreasing air recirculation.[12] These recommendations will obviously increase the energy consumption of HVAC systems, and it is as yet impossible to know how these health concerns will affect design standards or use of HVAC in the future.

Nevertheless, systemic shocks like the coronavirus pandemic and the climate emergency are challenging the viability of such sealed-facade buildings. See also note 7.02.

7.02: EXAMPLE

WHY SEALED BUILDINGS?

Increased concern about overheating due to climate change has favoured a mechanised fresh air supply. In addition, speculative projects, where end users are not involved at design stage, present particular challenges because specific tenant or local environmental priorities are poorly defined. However, sealed buildings cater for high-intensity use and all occupancy eventualities. They are often located on dense urban sites, where noise and pollution are also a risk. The result is standardised technical system design and a bias in favour of highly mechanised services strategies, towards a 'sealed' building (see figure 7.2.), particularly because energy costs are such a small proportion of business operating costs (figure 7.6).

Unrealised energy savings

Poorly commissioned and operated ventilation systems can compromise not only air quality but also energy efficiency. For instance, feedback studies in schools have found many instances where AHUs with inverters that allow variable-speed fans designed to adjust air flow and temperatures to changing conditions throughout the day were routinely not enabled. This meant that fans were constantly running at full power instead of at variable speeds, compromising the system's energy efficiency.[13]

Variable-speed fan-drive solutions can be set to respond to CO_2 levels so that fans vary their output to match demand, leading to energy savings. Fan variability needs to be enabled as part of the control strategy and checked during commissioning.

FIGURE 7.7: OPPOSITE
AHMM, White Collar Factory, Old Street, London, 2017. Interior view of 16-storey tower on a busy urban roundabout that features operable windows, a generous 4-m floor-to-floor height and exposed concrete soffits for thermal mass.

Natural ventilation: the ultimate low-energy ventilation system

It is important to determine at the design stage whether a building might be successfully naturally ventilated. The use of sealed facades is increasingly

FEEDBACK REVEALED THAT OFFICE WORKERS OPENED WINDOWS WHEN INTERNAL TEMPERATURES EXCEEDED 26°C, REGARDLESS OF ANY TRAINING RECEIVED

only acceptable where natural or mixed-mode ventilation approaches have been shown not to be able to provide comfortable indoor environments. This is also location and function dependent. See also note 7.03.

Naturally ventilated buildings do not rely on fans, pumps, AHUs or chillers to supply fresh air, which means there is less to go wrong. Instead, fresh air circulates through a building's spaces to reach occupants without the need for risers and ducts, relying on wind direction and pressure differences between inside and outside – see figure 7.7. This requires minimal or no energy use, making it the ideal low-energy ventilation solution.

However, the control of fresh air via natural ventilation can be compromised by human factors. For example, people tend to open windows too little when it is cold, which can lead to poor air quality, and too much when it is hot, exacerbating overheating.[15] Feedback revealed that office workers opened windows when internal temperatures exceeded 26°C, regardless of any training received – see case study 2.

In polluted areas, the amount that windows are opened needs careful attention. Outdoor pollutants are not detectable by smell, and the concentration of pollutants from traffic can increase where windows are open during warmer summer days for thermal comfort – see figure 7.8. Outdoor air may require filtering before it enters the building if mechanical ventilation is to be avoided.

Summertime night-purge ventilation, which high-thermal-mass buildings rely on for passive cooling, requires occupants to remember to leave windows open at night, and this often leads to the abandonment of regular

FIGURE 7.8

Keynsham Civic Centre, Bath and North East Somerset, UK. Air quality was monitored[14] – in summertime, when windows are mostly open, as shown by lower CO_2 levels, NO_2 levels rise and are much closer to outdoor levels. Here they are below WHO limits, but in polluted areas occupants should be prompted to close windows to prevent NO_2 build-up.

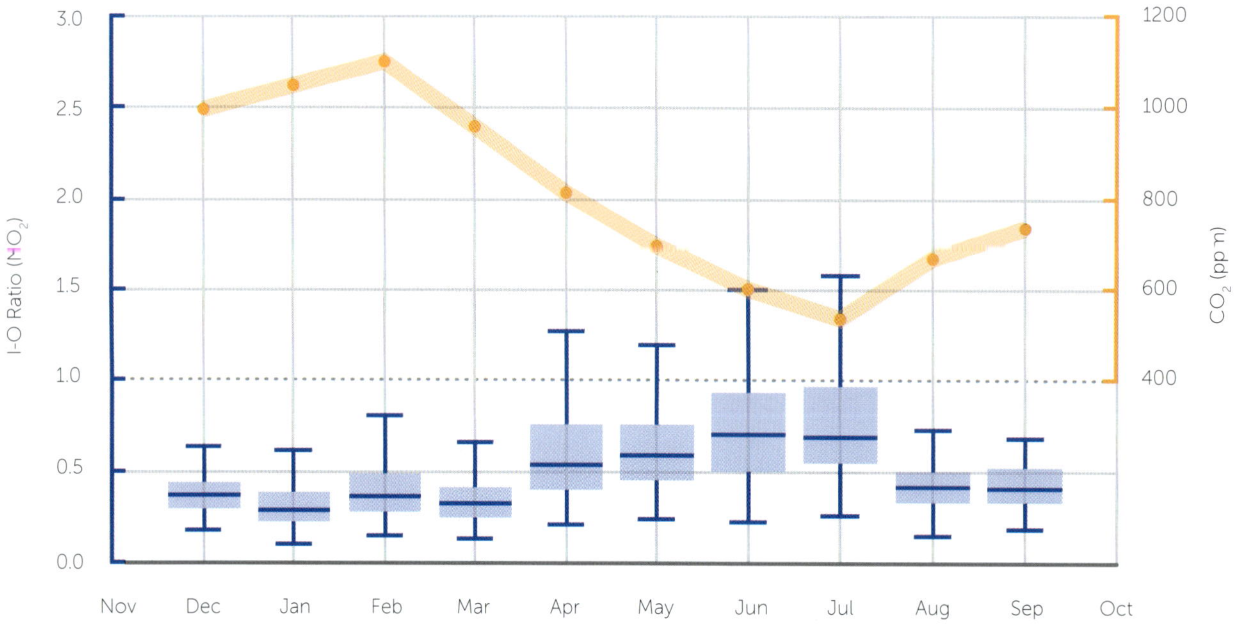

night-time cooling.[16] This abandonment is often due to security concerns, which can be overcome by installing security grilles on operable windows and positioning security sensors away from open windows.[17] However, these do not deter insects, and mosquito gauze greatly reduces air exchange volume. See also note 7.04.

In naturally ventilated buildings, the amount of fresh air that enters a building must be carefully balanced with summer heat gains and winter heat losses. As natural ventilation relies on natural forces, it can also be unreliable in providing good year-round air quality and thermal comfort.[18] For these reasons there has been an increase in low-volume continuous background ventilation (with or without heat recovery). Attention to heat recovery systems that minimise the risk of cross-contamination and air leakage is necessary to strike the right balance between the requirement for fresh air and energy efficiency. Increasingly, in most settings, window automation can be used to control air quality and temperature. Issues related to controls are discussed in chapter 8.

Mechanical ventilation with heat recovery

In buildings where fabric performance is outstanding, low-volume background mechanical ventilation with heat recovery (MVHR) can be sufficient to maintain thermal comfort and good air quality in wintertime. MVHR systems provide fresh air without losing heat from the exhausted stale air, and they are common in Passivhaus buildings. The heat is captured in the heat exchanger, so the fresh air intake never mixes with the extracted air[19] – see figure 7.9. In summer, MVHR is usually combined with natural purge ventilation.

The output of MVHR systems can be compromised by convoluted ducting that reduces air flow or requires higher fan outputs to achieve the desired air quality. Access to MVHR systems must be designed early on to allow for easy changing of air filters, and training must be provided for occupants to understand the controls and maintenance requirements to avoid pitfalls – see case study 3 and figure 7.9.

MVHR can operate in summer bypass mode so that heat from exhausted air is not recovered, avoiding bringing heat into spaces. Where the diurnal temperature changes are large (hot in the daytime and cool at night), MVHR in bypass mode delivers a form of background night cooling. And, when the outdoor temperature is colder than inside, MVHR also removes humidity from indoor air, increasing night-time thermal comfort in the summer. MVHR can be combined with heat pumps to pre-cool (or pre-heat) air, where demand is greater than that afforded by heat recovery only; this was for example the case in the Enterprise Centre (case study 7).

Labyrinths and earth tubes

The case studies in this book feature low-energy technologies for supplying fresh air and cooling. For example, the Everyman Theatre auditorium is cooled by earth tubes, a concrete underground labyrinth

7.03: EXAMPLE

NATURAL VENTILATION AND MIXED-MODE VENTILATION

The architects have replicated aspects of the natural ventilation approach from the Everyman Theatre (case study 1) on several other projects, including for Harvard University. However, when the temperature and humidity criteria are much more challenging, such as in US climates, it is not always possible to achieve full natural ventilation. This is where very efficient MVHR might be considered. Everyman Theatre head of facilities Robert Longthorne notes that 'natural ventilation in the auditorium has been highly effective and far cheaper to run than a conventional mechanical ventilation system'.

7.04: EXAMPLE

NATURAL VENTILATION AND USER INVOLVEMENT

At school 1, CO_2 monitors with a traffic-light system were specified in each classroom to prompt occupants to open and shut windows at the correct times. However, teachers found the system ambiguous in the busy classroom setting, and ended up ignoring it, leading to high internal CO_2 levels in winter and excessive internal temperatures in summer.

that cools fresh air to ground temperature and supplies it to the building via specially designed openings – case study 1. This technique can be used successfully to cool spaces, and can be incorporated architecturally – see for example figure 7.1. However, earth tubes are space- and material-hungry solutions that often cannot be implemented in dense urban locations.

FIGURE 7.9
Hamson Barron Smith, Carrowbreck Meadow, UK (2016). Diagram of MVHR system. MVHR units are located in easy-to-access cupboards, and filters can be changed without a ladder. User-friendly graphics explain how the system works, and a light flashes to indicate when the filter requires changing, with a reminder in biannual newsletters (case study 3).

1 Heat recovery ventilation unit

2 Fresh air vents

3 Extract air vents

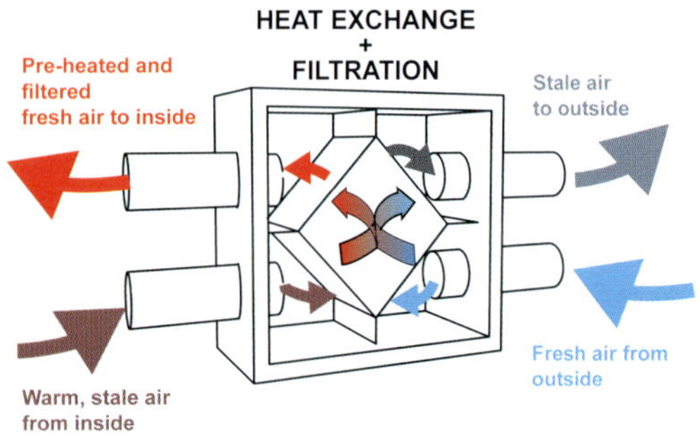

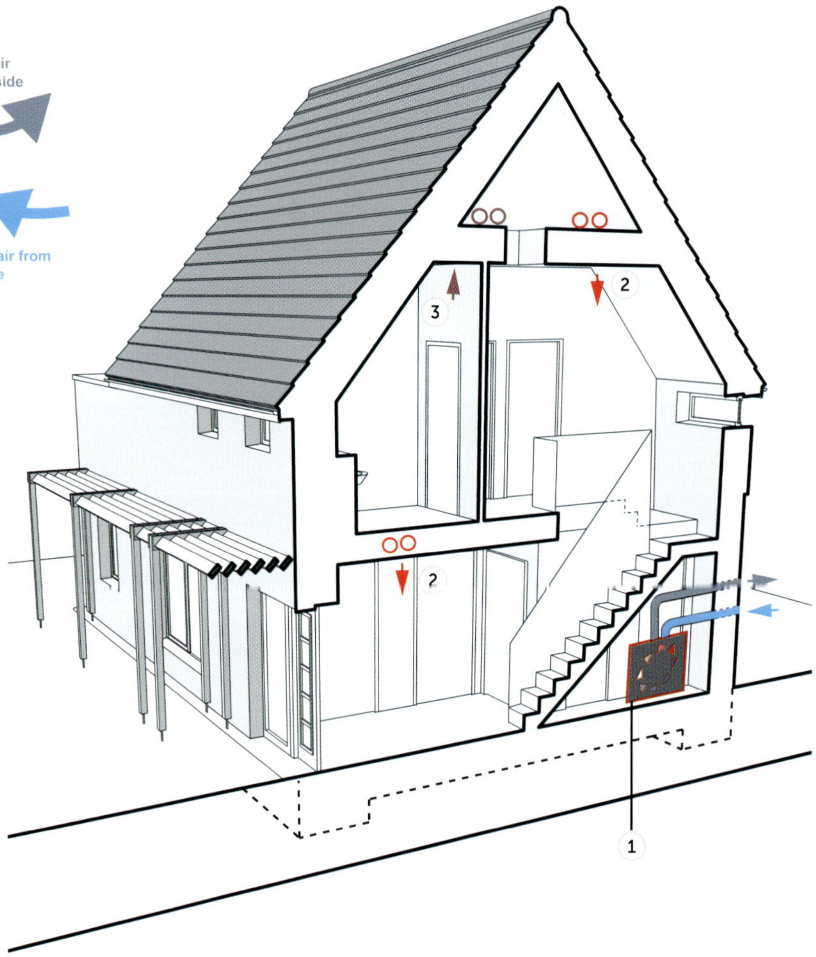

Heating and hot water systems

The simplest low-energy heating system is to use the body heat generated by building occupants. In super-insulated-buildings that maximise passive solar heating and rely on adaptive comfort, body heat and other internal heat gains can provide base space heating in many climates.

However, even when the need for space heating is minimised, most super-insulated buildings still need some form of low-carbon heating to bridge colder periods and for the provision of hot water (case study 4). Existing buildings with high space heating requirements should be made more energy efficient, prior to considering low-carbon heating systems. Buildings generally also need a 'back-up' system, to supplement the low-carbon heat when needed, or in case the main heat supply fails.

Specifying heating and hot water systems from the plethora of low-carbon heating technologies available can be daunting even for specialists. Comparing the effectiveness of different systems is not straightforward, and manufacturers' product literature can be overly optimistic about in-use performance. Ready availability of accurate information and data coming from feedback on in-use consumption and maintenance is paramount. This information would help design teams, facilities managers and end users to specify systems with more realistic efficiency expectations.

There are numerous potentially energy-efficient heating supply options, and some of their pitfalls are outlined below. These technologies are included because these are likely to become more prevalent in zero carbon buildings, as can already be seen in the case studies (chapter 9) and other examples throughout the book.

Heat pumps

In countries where the carbon intensity of the electricity grid is low, heat pumps have become a viable low-carbon solution for providing heating in climates where temperatures rarely go sub-zero for long periods. An important advantage of heat pumps is that they can provide low-carbon heating as well as cooling, as long as cooling demand is modulated by fabric efficiency such as solar shading. Feedback studies have shown the importance of specifying heat pumps that meet the heating demands of a building in tandem with the actual performance of the fabric and other services, to avoid the reliance on 'back-up' boilers, and their associated CO_2 emissions – see note 7.05.[20]

As with all novel technologies, feedback studies in buildings with heat pumps revealed numerous issues with correct installation, metering, commissioning and controls, which significantly affected their coefficient of performance (CoP).[21] However, where robust procedures are followed, they can offer a reliable and efficient alternative to traditional HVAC systems.

Solar water heaters

Solar water heaters (SWHs) are popular for their relative simplicity and low cost for both domestic and non-domestic projects. They can help meet both heating and hot water demand in buildings (case study 2).

The uptake of SWHs has been slowed in some countries by unfavourable utility rates, which can penalise their use. However, in many regions, especially warmer ones, they have become common. But, where the technology is still novel, commissioning problems are frequent. Systems are often found not connected and not metered, nor linked to the building management system (BMS), or not appropriately sized for the building (too big or too small) – see note 7.06.

Combined heat and power

Combined heat and power (CHP) uses gas or other fuels to generate electricity, and captures the waste heat for use on site. In order to be cost-effective and result in carbon emission reductions, a CHP plant only really makes sense where there is a high-enough year-round demand for heat.[24] See note 7.07

District heating

District heating is a more efficient way of providing heat than individually, building by building. Where it is sourced from waste heat or other low-carbon fuels, it can also reduce carbon emissions. The aggregate domestic hot water demand throughout the year in district or community heating systems is often high enough to warrant using CHP as a lead system. Consumption is normally submetered by end user for billing purposes (see chapter 3). Submetering provides the information to users to incentivise energy-efficient behaviour.

Biomass

Whole-life carbon studies of biomass installations have highlighted major concerns over the impact of growing, processing and transporting biomass fuel, as well as its biodiversity impacts.[25] Nevertheless, in some situations, biomass may still be viable.

Biomass installations require high-quality dry fuel, sourced locally, delivered in suitable quantities, stored in dry conditions and fed seamlessly into the boiler. These requirements can be challenging both in most urban contexts, and in rural environments. If the storage space is smaller than the standard delivery size, the cost of fuel shoots up. If the fuel absorbs too much humidity, or if the way in which the fuel is delivered into the boiler is disrupted, the boiler can fail. Biomass boilers also tend to require specialist operation, maintenance and repair. For example, at innovation centre 2, the annual maintenance budget for the biomass boiler was used up during the first month of operation.[26]

7.05: EXAMPLE

THE IMPORTANCE OF GETTING HEAT DEMAND RIGHT

Building performance evaluation studies revealed that the contribution of supplementary gas-fired boilers was significantly higher than expected at academy 3, where ground-source heat pumps were used as the lead heating system.[22]

7.06: EXAMPLE

SIZING SOLAR WATER HEATERS

At Bath University, the SWHs were oversized so they did not match the demand profile and had to dump excess heat in the summer. If properly specified and commissioned, SWHs can provide an excellent source of low-cost, low-carbon heat and hot water.[23]

7.07: EXAMPLE

CHP VIABILITY IS AFFECTED BY CARBON INTENSITY FACTORS

At the Everyman Theatre (case study 1), the CHP initially operated as intended, catering for the majority of the wintertime heat demand. Subsequently, supply chain issues affecting maintenance, and the simultaneous drop in the carbon intensity factor of grid electricity in the UK since 2018, has meant that the CHP has not provided the predicted carbon emission reductions.

Lighting

Indoor lighting has an enormous impact on occupant satisfaction, health and productivity (see chapter 2). Energy use from lighting can amount to a third of all energy used in a non-domestic building.[27] Hence it is necessary to reduce the over-reliance on artificial lighting by designing better daylit spaces (see chapter 5) and through improved lighting strategies.

ENERGY USE FROM LIGHTING CAN AMOUNT TO A THIRD OF ALL ENERGY USED IN A NON-DOMESTIC BUILDING.

One such potentially improved lighting strategy is the current move towards diffuse variable background lighting complemented by individual task lights, instead of the previous approach to mimic a uniform distribution of artificial daylight – see chapter 6. Moreover, the importance of finishes in influencing lighting conditions is increasingly recognised[28] and can affect the need for more or less lighting. For instance, at Keynsham Civic Centre

FIGURE 7.10
At school 1, light fixtures were positioned in rows perpendicular to the windows, so daylight dimming was curtailed.

(case study 6), the use of exposed cross-laminated timber (CLT) walls unintentionally lowered the lighting fixtures' lux levels as a result of less reflection compared to conventional finishes.

Moreover, energy-efficient lights such as LEDs help reduce energy use when artificial lighting is required. This is a fast-moving field that often requires specialist design. Products and systems types become obsolete quickly, and unforeseen technical challenges often occur when making new systems compatible with existing installations.

Nevertheless, the ability to supply light at variable colour temperature, intensity and shape with the use of minimal electrical energy has opened up vast opportunities. Because the wide availability of LED lights is relatively new, it is possible that unexpected problems might affect the quality of light or shorten the advertised life span of fixtures. Engaging an LED-trained designer and contractor early on ensures that the potential of LEDs can be fully exploited.

In addition, sensors and zonal controls that allow lights to adjust to occupant presence or absence, or daylight levels (i.e. daylight dimming), mean that energy consumption from lighting can be dramatically reduced. In order for daylight dimming to work properly, fixtures should be wired in parallel rather than perpendicular to windows, as illustrated in figure 7.10. Feedback has repeatedly shown that persistent issues with lighting controls mean that these benefits are often not realised.[29]

Equipment and appliance loads (unregulated energy use)

Equipment loads refer to the energy consumption of non-fixed building services and they contribute significantly to the performance gap.[30] This is because they are typically unregulated by building regulations, and therefore not included in building energy models (see chapter 4). This may be due to the fact that operating conditions and equipment installed by building occupiers are often not known at the design stage, and can vary widely depending on occupants (for example due to the absence or presence of large TV screens in housing). However, anticipating the energy consumption of equipment can be straightforward if likely patterns of use can be established through discussions with the client, estate manager and representative end users.

The impact of unregulated energy use on the performance gap can be exacerbated by the fact that the associated internal heat gains from the equipment are also not accounted for in energy models. Likewise, specialist equipment or functions that emit heat, such as server rooms, kitchens, trading floors in offices or science labs in schools, can significantly impact occupant comfort and energy use. These specialist functions are not considered in compliance calculations and excluded in the standardised calculation of a building's energy performance potential. Indeed, such specialist uses are often not planned for, and contribute to a performance gap once the building is in use.[31] This results in the underestimation of overheating potential in buildings with high heat gain equipment and high occupant density, and can lead to an increased demand for cooling – see note 7.08. Incorporating realistic appliance and equipment use in building energy models can help prevent the value engineering of passive and active cooling measures.

For unregulated loads, feedback is especially important – first, to monitor whether the building is performing as expected, and second, to base the building design on realistic assumptions of what the in-use unregulated load would be. While a high baseload from appliances can be detected from a building's daily energy profiles, a well-conducted users' survey will explain anomalies in metered data.

7.08: EXAMPLE

OVERHEATING EXACERBATED BY APPLIANCES

Omitting unregulated energy end uses can have long-term repercussions. For example, in another academy school, the research team's calculations showed that when actual appliance loads under realistic occupancy profiles were accounted for, classrooms were at a significant risk of overheating even under current climate conditions, and substantially more in future climate scenarios.[32] Similar findings forecast summertime overheating in low-energy dwellings.[33]

Design team engagement

Especially important to achieve zero carbon performance is understanding the implications of a building's services strategy on spatial configuration and envelope design, and vice versa. This is because a synergistic approach between spatial and fabric design can reduce the scope of technical systems. Moreover, technical systems must be carefully integrated into a building's overall spatial and envelope design, and with user experience (UX) in mind, in order to function effectively. They cannot be tacked on later; instead they must be proactively 'designed in' to a building from the start. Hence, multi-disciplinary scrutiny of technical systems is essential in ensuring zero carbon performance. Establishing a single party of responsibility in the design brief enables collaborative inter-disciplinary working that is needed to share assumptions early and work towards the same goals.

All systems have benefits and drawbacks, and a design team must match a building's services strategy to its context, likely end user needs and future adaptability. Getting the right balance between passive and mechanical systems requires an early assessment of likely operating conditions. This includes testing the resilience of design proposals against various operating scenarios to ensure that systems are sized with actual occupancy patterns in mind. One useful approach is to carry out a pre project survey followed by consultation with the client and end users to ascertain likely occupant density, hours of use, appliance use and affinity to adaptive comfort. Such iterative processes can bring actual performance closely in line with expectations are less immediately helpful in speculative projects, where the final tenants are unknown. However more feedback data has brought about a review of existing standards that are increasingly used by speculative projects too, see guidance in CIBSE TM 54 & 61-64.

It can be tempting for designers and installers to specify state-of-the-art equipment, when simpler systems would be equally appropriate. Similarly, planning conditions and sustainability certifications calling for on-site renewable energy provision can lead to token installations, and increased complexity. Because complex designs and novel technologies can lead to unforeseen problems, mapping potential risks with experienced technology providers must be part of a project's risk strategy. When deciding on system types, it is also important to plan ahead for expected maintenance needs.

Additionally, innovative systems require specialist expertise at every stage and should only be pursued if a client is committed to engaging the necessary specialists, and the required level of skill to specify, install, profile and manage systems is demonstrably available from design to operation. Including specialist commissioning in the employer's requirements and embedding it into the procurement process can dramatically improve the quality of installation, commissioning and effective use. Simplicity should be the default criteria for technical systems in any other case.

ACTIONS

PLAN

- **Assign responsibility for predicting, tracking and benchmarking energy performance and indoor environmental quality.**
- **Incorporate in engineers' scopes the carbon performance of systems, security of supply and energy efficiency, and risk assessment for air quality, taking into account human factors and user needs and experience.**
- **If a sealed building is considered an option, include in the project brief the appraisal of an operable facade approach.**
- **Include in consultants' scopes the calculation of whole-life energy and carbon costs at key RIBA project stages.**
- **Ensure involvement of the facilities manager (FM) and end users in developing services strategies.**

DESIGN

- **Evaluate options for a 'fabric first' approach with good year-round background ventilation.**
- **Assess opportunities for low-carbon provision of fresh air, cooling and heat.**
- **Incorporate in systems appraisal typical risks and a plan to address and manage these, especially with novel technologies.**
- **Design for adaptive thermal comfort in likely operating conditions, including the impact of occupancy patterns, hours of use, and appliance and equipment loads, and calculate energy budgets accordingly.**
- **Plan ahead for any installation and operational risks of low and zero carbon systems.**
- **Develop lighting design proposals bearing in mind human factors.**
- **Opt for simpler rather than state-of-the-art, complex solutions.**

Finally, irrespective of complexity, the whole-life cost of energy-consuming systems should be assessed for each project based on actual operational data. Systems design should take into account whole-life carbon, and bear in mind potential changes in the carbon factors of different fuel types.

PROCURE	USE	LEARN
• Provide a user-friendly summary of the services strategy and system types and their relationship to spatial and fabric design, including operating and maintenance requirements in employer's requirements and at handover. • Flag up in the project risk log any potential challeges associated with the installation and commissioning of building systems. • Follow Soft Landings guidance or another similar framework – see chapter 10.	• Engage occupants and train the FM in the correct operation of building systems. • Compare installed systems with specifications in employers' requirements.	• Summarise lessons learned and publish through professional platforms.

08

CONTROL SYSTEMS AND USER EXPERIENCE

FIGURE 8.1: OPPOSITE
Enterprise Centre, University of East Anglia, Norwich.

Introduction

Automation in buildings is controversial. Many see building systems digitisation as a panacea for underperformance, while others view it as one of the primary causes of the performance gap. The latter advocate minimising digitisation in favour of user control of passive design features to eliminate as much energy-consuming technology as possible from buildings. To ensure the best outcomes even for minimal automation, it is essential that controls operate as intended.

Control systems provide the interface between technical systems and:

- people (e.g. occupants opening or closing windows when they are prompted by a signal)
- the outdoor environment (e.g. windows automatically shutting when it rains)
- indoor comfort metrics (e.g. windows opening when indoor CO_2 concentrations or temperatures are too high)
- other technical systems (e.g. a back-up boiler kicks in when a primary low-carbon source cannot supply heat).

All interaction between the building systems, occupants and operators should be designed based on user experience (UX) design principles for a building to operate as intended. This is as much a design challenge as a procurement one, and this chapter focuses on both.

The simpler, the better

Control systems, in their simplest form, can be just on/off switches. Where they are placed, how they are labelled and how easy they are to operate can dramatically impact how effectively they are used. Frequently switches are placed out of reach of users, or without any indication of what they are for. And even with simple and manual controls, things can go wrong (see note 8.01). This is why a design guide for the specification, placement and labelling of switches is advisable.

Buildings that are larger and more complex, with more system types, zones and controls, may have a higher performance gap[1] – see figure 8.2. For example, when it is not clear what is controlled by a certain switch, users may inadvertently leave the switch 'on' or 'off', leading to poor indoor environmental quality and increased energy consumption.

Equally, how hard and for how long switches have to be pressed also affects how successfully people use the interface, with delays in response being especially disruptive in terms of how systems are used – see notes 8.01 and 8.02. Clearly, how intuitively occupants can interface with any animated part of a building is critical for successful operation.

Basic automation

In homes, the most common control system is a room or house thermostat, usually managing space heating with a pre-scheduled timetable to turn systems on and off. These systems often have interfaces that alienate users, with poor or inaccessible documentation making it difficult to change settings. In these situations, systems are left on when they are not needed, or in override mode, and the anticipated energy savings do not materialise. Feedback has documented cases of occupants who preferred to open windows to turn down the heating, instead of altering the complicated room thermostat.[2]

FIGURE 8.2

This chart shows the difference between predicted and actual carbon emissions of complex buildings, and shows a performance gap and higher-than-expected carbon emissions.[3]

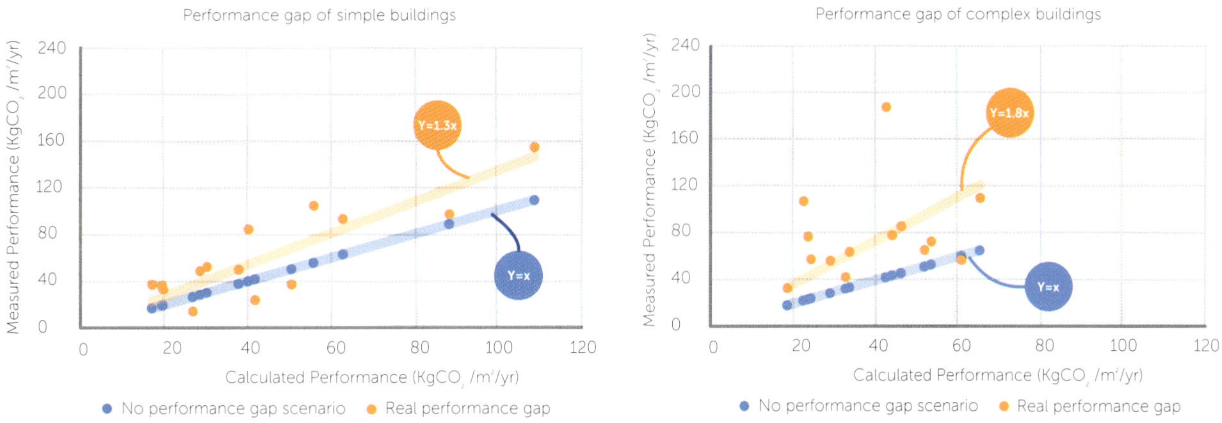

FIGURE 8.3
At school 1, fire doors were propped open to enable circulation throughout the day, compromising the low-carbon heating strategy across the entire school – see note 8.01.

8.01: EXAMPLE

GOOD DESIGN INTENTIONS WITH UNINTENTIONAL OUTCOMES

At school 1, feedback documented that fire doors, used by students and staff to move between wings of the school, were being left open all year round for the duration of the school day. This was because the caretakers did not realise that the manually operated latches could be retracted with a simple key to allow people to open and close the doors as needed. Instead, the fire doors were propped open (see figure 8.3), compromising the low-carbon heating strategy of the entire school. On cold days, the ground floor interior was uncomfortably cold, prompting the building manager to increase the heating flow temperature to 70°C. This meant that the expensive and efficient ground-source heat pumps (GSHPs), which have a maximum set point of 45°C, were never used, and the school ran on back-up boilers throughout the year.[4]

8.02: EXAMPLE

USER-FRIENDLINESS OF CONTROLS

Feedback from innovation centre 2 in Cornwall, UK, revealed a window switch with a 7-second delay instead of the 1-second delay specified. This led to windows behaving erratically when people repeatedly pressed the button to get a response, which meant that windows were not opened wide enough for summer cooling or narrow enough to admit a modest amount of fresh air in winter.[5]

8.03: EXAMPLE

SMART THERMOSTATS

'Smart' thermostats with improved interfaces that automatically account for variations in weather and occupancy use may yield improvements over conventional thermostats. Manufacturers claim that they save between 10% and 12% on heating costs, and 14% on cooling costs, though this may be overestimated.[6] See also note 8.17.

Building management systems (BMS)

Automated controls allow technical systems to vary their operating times and intensity to match occupant demand through a building management system (BMS), which modulates or switches equipment on and off based on a pre-programmed schedule, allowing for example a heating system to come on in the winter at 6 am and switch off at 4 pm. A BMS uses the same type of internet protocol (IP) as a computer network to control the operation of different technical systems, such as a boiler, air-handling units (AHUs) and so on.

A BMS can control both passive systems, such as window opening, and active systems, such as switching AHUs on and off. It can control one specific end use in a building, such as fixed lighting or security systems, or multiple systems, including heating, ventilation and cooling, security, vertical transportation, or fire alarms. Increasingly, in larger buildings, a BMS manages all a building's systems, ideally through a single digital interface. Building regulations tend to incentivise specifying a BMS to demonstrate that technical systems would only operate when needed.

A BMS turns systems on and off or modulates outputs, during a scheduled period, by checking sensor readings through a physical or wireless connection. If conditions reach a setpoint the BMS triggers an action. For example, at the Rocky Mountain Institute (see case study 4) the external blinds are activated when the concrete slab reaches a certain temperature. A BMS also communicates performance and errors to personnel on-site or remotely to effective and timely troubleshooting – see note 8.04.

Successful BMS operation requires that every element works correctly.

- The sensors in technical systems must be properly installed, calibrated and connected to the BMS.
- The systems must be correctly profiled so they are programmed to tailor outputs to set points and schedules that reflect varying occupant needs.
- The data from sensors and meters must be logged in the BMS memory for diagnostics.
- System performance and errors must be communicated in an accessible format on-site and, if possible, remotely.

Theoretically, BMS offer significant benefits to building owners and operators, as well as tenants, providing good indoor environmental quality (IEQ) with increased energy efficiency. On one test building a BMS was shown to reduce energy consumption by over 30%.[7] However, the complexity of BMS (see figure 8.4) means that they require proper and competent configuration and use. Feedback data has shown that poorly configured BMS can actually increase energy use, compromise IEQ and increase costs.[8] Shockingly, poorly configured BMS account for as much as 8% of the UK's total energy use.[9]

Typical BMS issues

Most building automation problems stem from poorly configured sensors and BMS settings, but they may also be due to incorrect zoning or faulty metering. Often these issues are only discovered when building performance monitoring shows up unexpected energy use or underperforming systems. A feedback process can be the starting point for troubleshooting and, ultimately, remedial action.

8.04: EXAMPLE

BMS AND EFFECTIVE TROUBLESHOOTING

At the Enterprise Centre (see case study 7), the BMS was set to automatically email both the project architect and services engineer throughout the three-year Soft Landings period if temperature or CO_2 surpassed certain levels. This ensured prompt troubleshooting, including an instance where a lightning strike had disabled all the temperature sensors.

POORLY CONFIGURED BMS ACCOUNT FOR AS MUCH AS 8% OF THE UK'S TOTAL ENERGY USE.

FIGURE 8.4

Keynsham Civic Centre (case study 6) – BMS snapshot showing the complexity of and interactions between different systems.

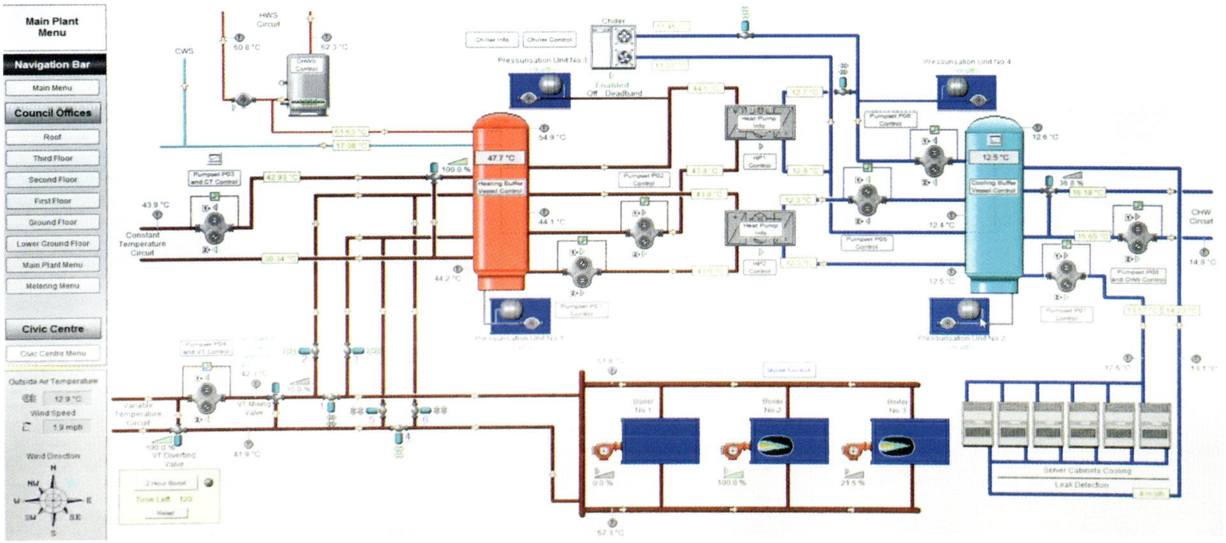

8.05: EXAMPLE

IMPORTANCE OF ZONING OF LIGHTING

At school 1 the controls were designed so that lighting in different corridors would switch on when occupants triggered motion sensors. However, during commissioning, all corridors were assigned to the same zone. This meant that all the corridor lights went on whenever a motion sensor was triggered in any of the corridors.[10]

Zoning and intermittent occupancy

Feedback studies repeatedly find that poorly planned functional and thermal zoning – the matching of spatial layouts to likely needs – leads to excessive energy use[11] – see chapter 5. Likewise, significantly increased energy use is often found in buildings that accommodate intermittent and out-of-hours activities, in particular educational buildings during unoccupied hours and holidays. This is because controls are often unable, or are not programmed, to reduce the operation of technical systems to specific locations or intensity. This is usually due to sensors and equipment not being assigned to the intended zone, or because individual zones are not correctly profiled or not profiled at all – see figure 8.5 and note 8.05. For example, heating systems that have different profiles, such as underfloor heating or radiators, should be located as well as programmed as different thermal zones to allow for independent control.

Where partial or intermittent occupancy is common, systems should be scheduled so that only the occupied areas of a building are operated – see chapter 5. This must be anticipated during design by discussing anticipated out-of-hours use with the client, and if possible with end users and the building manager or facilities manager (FM).

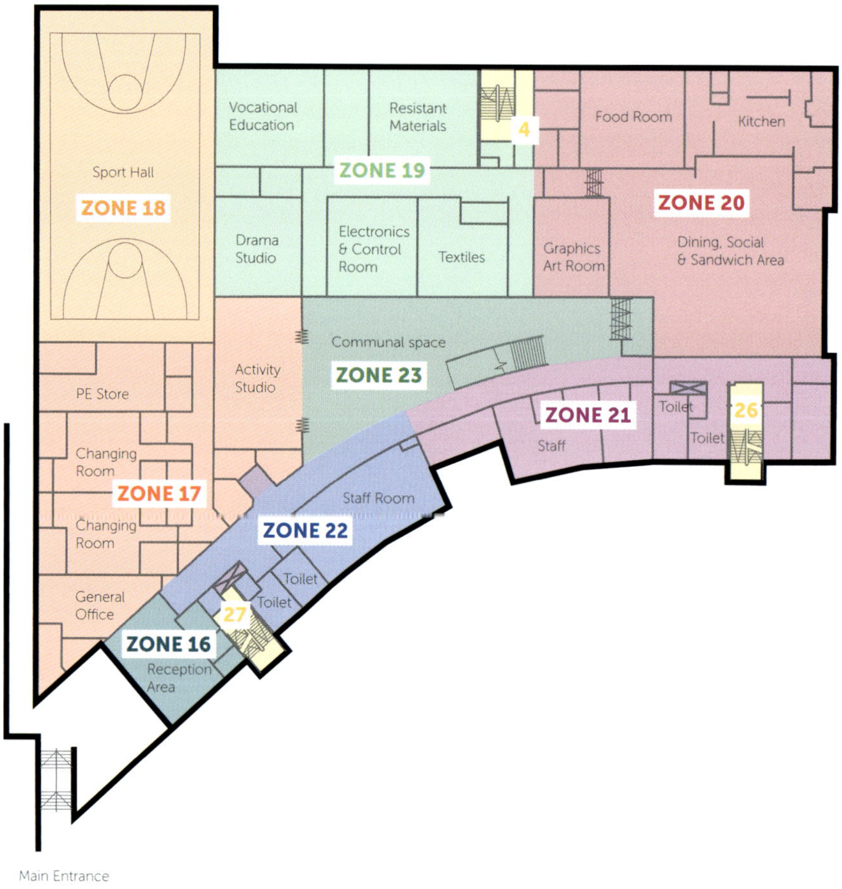

FIGURE 8.5

Academy 3 zoning diagram. The coloured organisational zones were not reflected in the BMS profile, so the building manager was unable to switch systems in specific zones on or off when certain spaces were used out-of-hours. As a result, energy performance could not be optimised.

Sensors and sensor locations

Sensors are the fundamental building blocks of a BMS, and understanding what they do and where they should be located can help you avoid common pitfalls downstream. Sensors can detect on/off states and absence/presence (Y/N), and they typically measure output, room temperature, CO_2 levels, relative humidity, lighting levels (lux) and sound levels (dB). With growing concern about health and wellbeing of occupants in buildings, volatile organic compounds (VOCs) and particulates are increasingly being measured.

Feedback consistently reveals problems due to disabled, incorrectly calibrated or poorly located sensors that significantly impact human comfort and energy use – see note 8.06. Problems with CO_2 sensors can result in either too much or too little ventilation, leaving occupants too cold or drowsy.

Sensors sometimes communicate directly with end users, and if so they must have a user interface with clear and unequivocal information. If not, people can unintentionally do the opposite to what is required – see notes 8.07 and 8.12.

Finally, inappropriate sensor placement, such as locating sensors in spaces which are cooler or warmer than other spaces, leads to overheated or underheated spaces elsewhere – see note 8.08.

Set points and system profiles

The output of technical systems is regulated by set points and profiles, set up in a BMS. If these are incorrectly set, IEQ and energy consumption can suffer.

For example, the naturally ventilated classrooms at school 1 rely on CO_2 as well as temperature sensors to open and shut plenum vents for the cross-ventilation of classrooms[12] – see figure 8.6. If temperatures reach 24°C in summer, overhead plenums are programmed to open fully. In winter, when CO_2 levels reach 1200 ppm, the plenums are programmed to open marginally and then close as classroom temperature drops to limit heat loss.

However, the temperature sensors were never enabled during commissioning, and the system operated on CO_2 triggers only, causing a range of knock-on effects. In summer the classrooms overheated because as soon as the windows opened, CO_2 levels dropped and the openings automatically shut, compromising the cross-ventilation and causing temperatures to soar. In wintertime the plenums were constantly open as teachers did not open the windows due to the confusing prompts from CO_2 sensors – see note 8.07. Cold drafts contributed to the heating flow temperature being raised to 70°C.[13] This increased the school's actual carbon emissions four-fold compared to design estimates. Missing documentation meant that maintenance staff struggled to diagnose and fix the problem.

8.06: EXAMPLE

SENSOR LINKING PROBLEMS

Feedback undertaken at Oxford Brookes University found that the on/off status of temperature sensors in the hot water tanks in one buildingwas not linked to the BMS, causing immersion heaters to be permanently on.[14] A similar linking issue was found at innovation centre 2, where the boiler's low-fuel sensors were not connected to the BMS, triggering the back-up boilers.[15]

8.07: EXAMPLE

USER INTERACTIONS AFFECT PERFORMANCE

Sometimes occupants or operators manually override certain systems without fully understanding the overall strategy. For example, CO_2 sensors at school 1 prompted teachers to open or shut windows depending on CO_2 levels in a classroom. Due to the confusing interface, teachers ignored the signals, and classroom CO_2 levels frequently rose to unhealthy levels throughout the academic year.[16] At another building, automated exterior sunshades ended up being used to control glare rather than heat gain, shading the interior far more than originally intended.

8.08: EXAMPLE

IMPACT OF ISSUES WITH SENSORS

At academy 2 in London, the sensor for the cooling system was incorrectly placed in the IT room, which was always several degrees warmer than the rest of the school. When the IT room temperature exceeded 18°C, the school-wide cooling system switched on, fighting the heating system all year round. Occupants resorted to portable heaters in winter, and the heating often switched on in summer. Energy bills were more than double predicted estimates.[17]

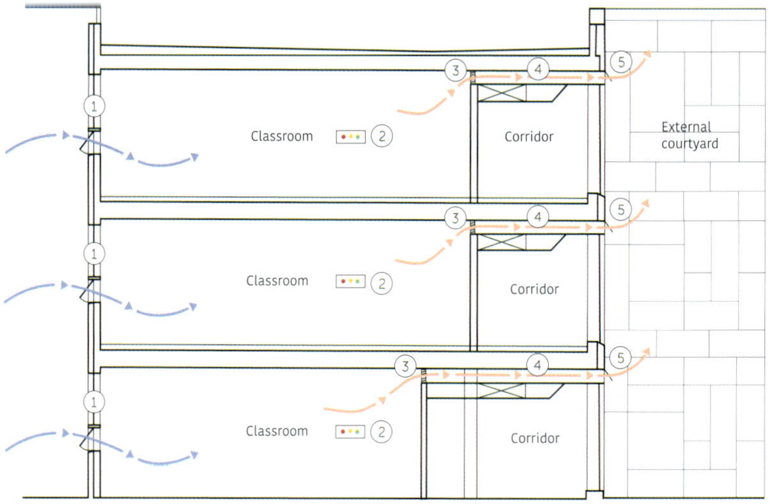

FIGURE 8.6
School 1 features a plenum natural cross-ventilation strategy that relies on a combination of manual and automated controls.

8.09: EXAMPLE

METERING STRATEGIES

In the UK the provision of a submetering strategy in a building logbook and the testing of meters during commissioning are regulatory requirements on all non-domestic buildings. For example, at 201 Bishopsgate, electricity consumption was reduced by 20% by good submetering.[18]

8.10: EXAMPLE

LOW CARBON TECHNOLOGY METERING

Low-carbon technologies should be separately metered and monitored to determine if they perform as anticipated, and to understand their contribution to energy savings. For example, heat pumps will have separate meters for electricity used and heat generated.

The importance of (sub)metering

A building's total electricity and fuel consumption is the sum of the measurements taken by the main meters and any energy generated and used on site. Metering facilitates energy management and provides the energy supplier with the basis for charging for energy used. Effective metering is essential to accurately establish a building's energy profile, yet feedback studies are constantly hindered by faulty or inadequate metering.

Establishing a metering strategy at design stage is an essential step towards validating performance in use – see note 8.09. Ideally, each end use in each thermal zone should be separately submetered – see also chapters 3 and 6. Any problems with meters impede energy management, because if readings cannot be trusted then it is difficult to ascertain if certain equipment, an end use, a zone or even a whole building is functioning as intended. Moreover, just reporting energy performance data is not enough.

The total readings from submeters should add up to the total consumption of fuel and electricity in a building (i.e. meter reconciliation – see chapter 3). Usually, the more complex a building, the more complicated its metering strategy, so establishing the metering strategy at design stage is an essential step towards validating performance in use. Ideally, each end use in each thermal zone should be separately submetered. Meters are expensive, and the cost for a large non-domestic building can amount to tens of thousands of pounds or euros. Therefore any investment in metering must effectively support the reconciliation of energy end-use consumption with the main meter readings, to identify excessive consumption and establish the carbon savings attributable to energy-efficient building fabric and systems, and low and zero carbon technologies – see note 8.10.

Seasonal readings must also be checked against performance expectations, so that issues can be diagnosed and rectified – an example of this is in note 8.11. Any errors can lead to lack of reconciliation of energy use, adding uncertainty to all submeter readings. Meters should be installed, calibrated and connected to a BMS which automatically logs meter readings to enable the automatic monitoring of building energy consumption trends. Manual reading of meters is a lengthy, painstaking and costly process, which should be a last resort.

Feedback highlights major design opportunities to improve the functionality and the use experience (UX) of sensors, system interfaces and metering.

Communication between systems and people

Large, complex buildings increasingly rely on automated technical systems and controls to achieve good IEQ for occupants. Despite this increasingly complex technology and automation in buildings, people remain central to

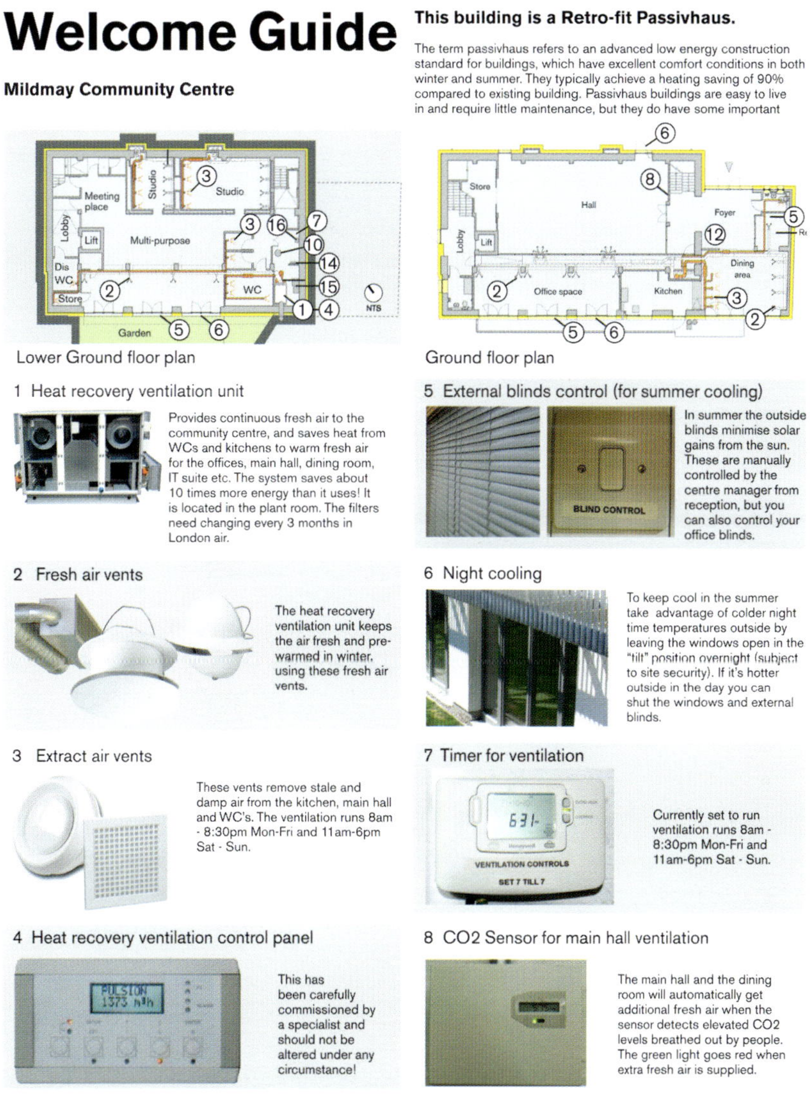

FEEDBACK HIGHLIGHTS MAJOR DESIGN OPPORTUNITIES TO IMPROVE THE FUNCTIONALITY AND THE USE EXPERIENCE (UX) OF SENSORS, SYSTEM INTERFACES AND METERING.

8.11. EXAMPLE

SENSOR DATA OUTPUTS

At the Enterprise Centre (see case study 7), the architects set up simple templates for data in order to get simple outputs, and this helps them interrogate the data and what it means. Architect Gareth Selby says: 'For example, one room is 2°C hotter than another. Then ask yourself, "Is that good or bad? What needs to change?" This then clarifies what to look for in terms of issues and informs the survey questions that you ask people.

FIGURE 8.7

Building user manual for Mayville Community Centre, London by bere:architects – an excellent example of clear communication of a control strategy to occupants and the FM.

ensuring that buildings perform as intended. Good building performance relies on good communication – between systems, but also between systems and the right stakeholders – see figure 8.7. Indeed, lack of clear communication has been highlighted as a major obstacle to good building performance.

Where any automation is present, it is especially important that the channels of communication between different systems, and between controls, installers, operators and occupants, all work smoothly – see figure 8.8. The BMS connects to and controls the building services sensors following internet protocols (ideally via a dedicated broadband channel). External access and connectivity to the BMS must be completely secure, for remote access, monitoring, inspection and data collection. Ideally a single BMS should control all services to avoid conflicts (for instance if automated ventilation and heating are on a separate BMS).

If the commissioning of building services and systems is compromised, it can have costly impacts on performance and on the cost of undertaking building performance evaluations (BPEs) – see note 8.13. While undertaking a BPE would normally cost a conservative 0.1% of a construction budget, when the correct information needs to be tracked down due to poor commissioning

FIGURE 8.8
Diagram showing the complexity of communication between technical systems and controls, and controls and end users.

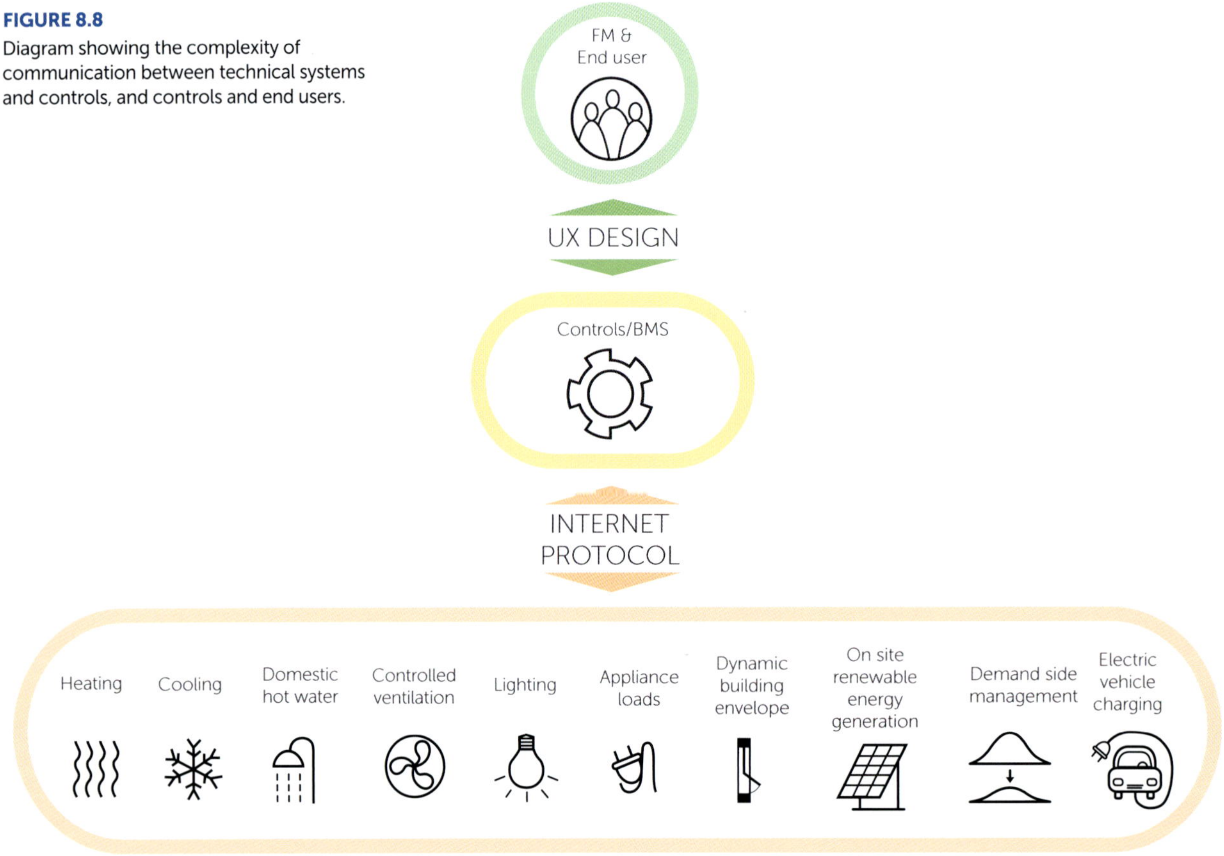

or communication, the cost of a BPE can at least triple.[19] What is also clear is that without feedback, these problems would not be identified.

Systems need to be properly connected, calibrated and profiled for the BMS to balance output from each system to match demand – see note 8.12. Building studies have demonstrated that issues with controls can compromise the best services strategies, and that the majority of these relate to poor user experience (UX) design or procurement.

Procurement

Ideally, all project stakeholders should have input into the design and delivery of controls, as they are critical to the healthy functioning of a building and its systems. At different project stages responsibility for the successful integration of controls will shift to different project partners – see figure 8.9. That means that accountability must be transferred smoothly. The intended operational settings of the building must be specified and communicated to installers, and their correct implementation must be recorded by the installers and communicated to building operators. To avoid commissioning errors, it is important that building operators are appointed prior to commissioning, and that

8.12: EXAMPLE

CONFLICTING CONTROL SYSTEMS

At innovation centre 1, the heating system was controlled by the BMS and automated windows were managed remotely by a different system, with miscommunication between the different control systems. As a result, when room temperature dropped below a certain threshold, heating increased, which inadvertently caused the automated windows to open to cool the room. As a result, occupants were frustrated by their inability to control the systems better and energy was wasted.[20]

8.13: EXAMPLE:

COMMISSIONING THE BMS

BMS software needs to be connected to a building's IT network and ideally to the internet too – i.e., it should be installed after the IT network is fully operational. Commissioning must include all low and zero carbon systems, usually supplied by specialist subcontractors. These are usually the last systems to be installed, squeezing commissioning time. Soft Landings offers a helpful framework for avoiding common pitfalls, and requiring main contractors to bid for building contracts with a named mechanical, electrical and plumbing (MEP) subcontractor can also overcome some of these obstacles.

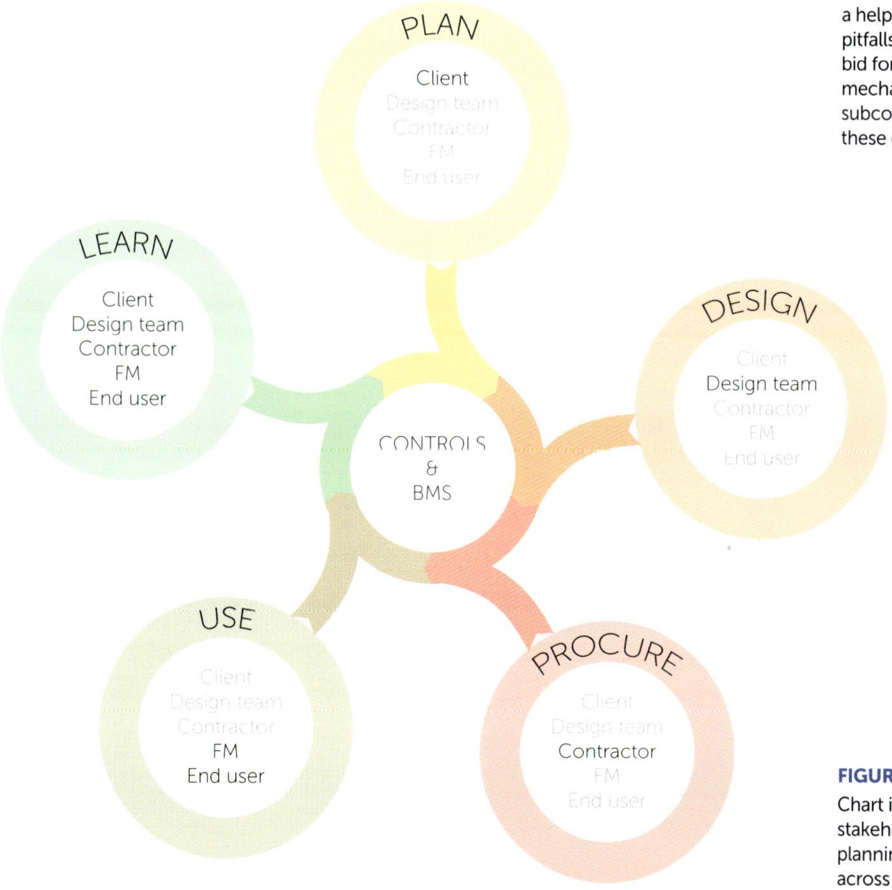

FIGURE 8.9

Chart illustrating the involvement of different stakeholders in control systems and BMS planning, design, procurement and use across project stages.

handover procedures are followed in full, allowing enough time – see note 8.13. Transferring the control strategy to the BMS during commissioning represents the highest risk, and timing is critical.

BMS and user experience (UX) design

Complex and confusing user interfaces make it difficult for building managers (or users) to change specific environmental settings or collect performance data. The interface to specific services should be informative and intuitive, and training should not be so onerous that there are problems retaining in-house knowledge. Documentation should also be clear, relevant and to the point, and alarms and error messages should be easy to manage, otherwise maintenance staff will be overwhelmed. Permission levels for end users should be flexible enough to allow changes to all parameters described by the building's control strategy – see note 8.14.

Access is also a physical challenge: the BMS interface should be accessible online via a single interface, not on different computers in remote or uncomfortable spaces – see note 8.15.

Any interaction between occupants and building systems is a design challenge. As demonstrated by some of the case studies, air-quality sensors should invite correct user behaviour intuitively. This is a design opportunity yet to be resolved between architects, product designers and suppliers. A key purpose of smart buildings is for systems to be able to self-adjust to occupant behaviour, as further explored below.

The future of BMS: automating dynamic change

As technology advances, new, better sensors can be integrated into more services, with the ability both to provide more real-time data and to affect a 'smart' change in the service provided. Such a smart approach requires minimal human intervention and can help systems run more efficiently.

A smart system is intended to adapt operations in response to variations in occupancy and demand in order to maintain comfortable conditions with minimal energy consumption. The smart system actively monitors real needs and usage, and 'learns' from occupants and operators, actively adjusting schedules, set points and other control functions in response to changing conditions (such as occupancy patterns or weather), rather than completely depending on a pre-set schedule – see notes 8.16. and 8.17. A smart system can also detect malfunctioning equipment. For example, if a window actuator jams, the system could message the person responsible with the nature and location of the error.

The vision for smart systems is to be able to process feedback between sensors, people, building fabric, energy sources and controls without requiring intervention by specialists. Clearly this could transform building management, enabling a highly effective, proactive energy-management approach – with major benefits for low-energy building performance. The longevity and 'future-proofing' of these smart systems is yet to be determined, but could

8.14: EXAMPLE

BMS PERMISSIONS

Reduced BMS access levels at school 1 meant that missing temperature set points for classrooms could not be added by the FM because rights to the BMS underlying dependencies were limited to the original installer.[21]

8.15: EXAMPLE

BMS ACCESS

At innovation centre 2, the security system, access control, BMS and fire alarms were all located on different computers, installed by separate subcontractors.[22]

8.16: EXAMPLE

SMART CONTROLS

A smart control system in an office building would log if a space that is scheduled to be heated or cooled at a certain time is in fact unoccupied, and it would turn the heating down or switch the cooling off. If this happened regularly at a certain time, the smart control system would update the room's heating schedule accordingly.

8.17: EXAMPLE

SMART HOMES

In domestic settings, wifi-connected digital controls – or increasingly mobile devices (phones, watches, etc.) create new opportunities for smart systems to respond to variable environmental conditions and occupant needs. Such systems could be automated to prompt people to open/close doors and windows or external shutters. These systems can bring significant quality-of-life benefits, particularly for the elderly.

be supported by a quality assurance scheme – see note 8.18. For example, issues of compatibility between sensors, software, and building systems across the years have not been tested. Furthermore, the resilience to loss of power or malicious tampering of smart systems has yet to be assessed. A second issue is that these smart systems collect a large amount of data, raising concerns about privacy, though these concerns might be minor for public buildings, where any activity is by definition 'in public'.

Smart Readiness Indicators (SRIs)

Voluntary 'Smart Readiness Indicators' (SRIs) were introduced in the EU in 2020 as part of the Energy Performance of Buildings Directive (EPBD),[23] and are intended to evaluate how smart a building is in terms of automation and monitoring of technical systems. In addition to the seven energy end uses previously described (heating, cooling, hot water, ventilation, lighting, automation and monitoring) and that are provided for by technical systems, SRIs add three additional domains, providing strong commercial drivers for smart systems: on-site renewable energy generation and storage, demand-side management (DSM) and electric vehicle charging.[24] To qualify as 'smart', each domain must meet eight impact criteria:[25]

- Energy saving on site
- Flexibility for grid and storage
- On-site generation
- Comfort
- Convenience
- Supporting wellbeing and health
- Maintenance and fault prediction
- Providing information to occupants

SRI evaluation is still based on a building's assumed, rather than actual, capability after installation. In line with the premise of this book, a more useful way to evaluate a smart building would be to analyse the in-use data that smart systems constantly collect and rely on, validating a building's performance and the purpose of the EPBD.

When it comes to achieving occupant satisfaction and comfort in buildings at zero energy performance, some industry professionals regard smart building systems as a panacea, while others see them as yet more expensive technology. While the benefits are compelling, smart systems add a further layer of complexity to technical systems and require specialist expertise and supply chain integration for their design, installation and maintenance to make the most of their promise.

Designers would be well advised to understand these novel technologies if they want to successfully integrate them into building design. Accurately designed and metered smart buildings not only ensure that the automated systems operate on the basis of robust data, but are well placed to share data for the purposes of better industry benchmarking – an essential step in the journey to zero carbon buildings.

8.18: EXAMPLE

REGULATION/QUALITY ASSURANCE OF SMART SYSTEMS

Smart systems could make buildings more responsive to people's needs, easier to operate and more efficient, while also facilitating performance monitoring. However, smart control systems are usually excluded from building legislation – a regulated independent quality assurance scheme for data reconciliation is needed to contribute to benchmarking and help close the performance gap.

Reconciliation of performance

The validation of a system's performance is an essential component of building commissioning and handover. However, due to the complexity of the process and the fragmentation of responsibilities, a thorough performance reconciliation rarely takes place. This means that in practice most control systems encounter problems from the very beginning of operation.

As such, a 'reconciliation' quality assurance scheme, via independent validation, would be needed to ensure that systems and controls are

ACTIONS	PLAN	DESIGN
	• Set realistic 'smart' building targets. • Undertake sufficient consultation with end users to establish the most resilient zoning in a building. • Understand likely occupancy patterns; discuss with building management what parts of a building are likely to be occupied, how intensively and when they are/are not in use. • Ensure that zoning of the services matches out-of-hours use patterns. • Make out-of-hours use explicit in the controls strategy; discuss with users and check expectations. • Ensure smart systems are planned for gathering performance data after installation.	• Specify appropriate thermal and lighting zones, systems and programming for successful out-of-hours scheduling. • Match set points and profiles to occupant needs. These settings need to be to be purposely added based on evidence and full information. • Ensure that controls and their interfaces are designed to be user-friendly. • Ensure smart systems are designed to gather performance data after installation. • Evaluate proposed smart system design according to EU Smart Readiness Indicators (SRIs).

designed, installed and enabled for their full intended purpose, 'reconciling' their expected performance with their actual performance. Effective management of building systems and meaningful feedback are only possible once these measurements, readings and reports match the key performance indicators (KPIs) set out in the controls strategy. This type of evaluation could make Energy Performance Certificates more robust, and would be an important first step in guaranteeing that a building meets its expected energy performance in the real world.

IN PRACTICE MOST CONTROL SYSTEMS ENCOUNTER PROBLEMS FROM THE VERY BEGINNING OF OPERATION

PROCURE	USE	LEARN
• remote reporting of energy end use and IEQ indicators in control system specification • Require monthly reporting from the first month of occupation so that problems can be identified and rectified during the defects liability period. • Retain an adequate bonus until all issues are rectified. • Ensure data protection is in place according to the highest standards and legal requirements. • Ensure that correct targets and incentives are incorporated into the contractor's prelims. • Ensure that subcontractors have the necessary expertise, incentives and information. • Ensure that zones are properly defined in the controls strategy and are clearly communicated in work packages and building manuals. • Ensure that each energy end use is submetered and each thermal zone has its own set of submeters for each energy end use.	• Ensure that sensors are properly located, enabled and calibrated. • During commissioning, check all meters are in place, enabled, calibrated and connected to the BMS. • Ensure BMS settings accurately reflect occupant needs and are not just left at default settings. • Seasonally commission and reconcile predicted versus measured energy use, and undertake independent validation of data reconciliation. • Utilise smart systems data to monitor performance remotely and cost-effectively. • Ensure that when performance issues occur, they are diagnosed and fixed. • Highlight to building managers and users that if use patterns change the BMS should be updated to match. • Confirm that the sum of submeter readings matches the total energy and fuel consumed. • Ensure that BMS data is collected and stored long term – ensuring data protection protocols are followed.	• Documentation should be made available to all of the design team, and the lessons learned shared both internally and publicly.

everyman

09

CASE STUDIES

Introduction

Emergent understanding of the important role of feedback in delivering low-energy buildings means that more and more performance evaluation is underway. Nonetheless, the number of projects with data available for publication remains low.

The buildings in this book were chosen from the limited pool of exemplary projects with ambitious performance targets and reliable data because they represent a variety of building types: offices, higher education, a school, housing and a theatre. Two projects extend the geographic reach of examples beyond the UK: one in Denmark and one in America. Most feature natural ventilation in some form.

Ideally feedback is planned for from the start, with 'before' and 'after' occupant surveys, as well as an agreed strategy for monitoring energy and indoor environment quality (IEQ) data. Of the seven buildings, three had elements of feedback contractually agreed: the Rocky Mountain Institute (case study 4), Keynsham Civic Centre (case study 6), and the Enterprise Centre (case study 7). Moorlands Rooftop (case study 2), an architectural practice's own offices, was used by the architects to investigate many of the principles espoused in this book. At Carrowbreck Meadow (case study 3), the architects committed to the Soft Landings framework at project inception. At Frederiksbjerg School (case study 5), post-occupancy interventions tested the impact of different lighting solutions on student concentration and energy efficiency. Finally, the Everyman Theatre (case study 1) benefited from the involvement of an engaged head of facilities from the outset.

In contrast with many architectural publications that focus on the 'latest' built projects, buildings with performance data, by definition, are not brand new. The buildings featured in this book were completed between 2014 and 2016. It often takes two to three seasons for technical systems and controls to be fine-tuned for optimal performance so data can be collected and analysed and issues rectified. This initial bedding-in period is only a fraction of a building's lifetime. To be truly sustainable, buildings must be built to last, and cherished and valued by their occupants so they are well cared for over the years.

The buildings shown on the following pages exemplify inspirational design informed by climatic intelligence – what Dean Hawkes describes as 'environmental imagination'.[1]

These seven buildings demonstrate that it is possible to design, construct and operate buildings differently in the face of the current climate crisis. And this is not only possible, it's essential.

CASE STUDY 1
EVERYMAN THEATRE

TYPE NATURALLY VENTILATED THEATRE

LOCATION LIVERPOOL, MERSEYSIDE

COMPLETION 2014

CLIENT LIVERPOOL AND MERSEYSIDE THEATRES TRUST

GROSS FLOOR AREA 4690 M^2

COST £13.4 MILLION

COST PER SQUARE METRE £2,300

SERVICES COST £500 M^2

EXTERNAL WORKS 300 M^2

NUMBER OF OCCUPANTS AUDITORIUM 400 + 40 STAFF AND ACTORS

SERVICES GAS BOILERS, AIR-SOURCE HEAT PUMPS AND CHP, NATURAL VENTILATION VIA PLENUM BELOW AUDITORIUM

AIRTIGHTNESS 5 M^3/HR/M^2 @ 50 PA

FORM OF CONTRACT JCT STANDARD CONTRACT

ENVIRONMENTAL CERTFICATION BREEAM EXCELLENT

ENVIRONMENTAL RATING DEC B49 (2016-17)

ARCHITECT HAWORTH TOMPKINS

SERVICES ENGINEER WATERMANS BUILDING SERVICES

STRUCTURAL ENGINEER ALAN BAXTER & ASSOCIATES

CONTRACTOR GILBERT ASH

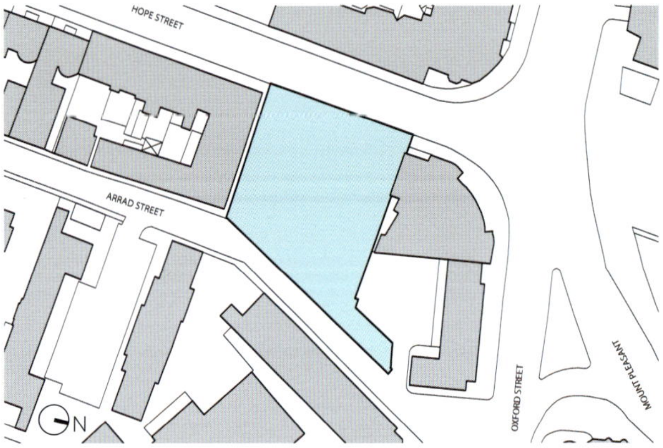

FIGURE 9.1.1
Location plan.

FIGURE 9.1.2: OPPOSITE
The four brick chimneys clearly express the theatre's natural ventilation strategy.

The story of how the RIBA Stirling Prize-winning Everyman Theatre came to be a low-energy, naturally ventilated building on a city centre site in Liverpool is one of serendipity – with a visionary client willing to take a risk, architects with a deep intuitive understanding of how to create the best theatrical experience, and a services engineer determined to make it work with reduced energy needs.

The Everyman Theatre was completely rebuilt between 2011 and 2014 after a long history in Liverpool, which had cemented its position as a cultural icon in the minds of Liverpudlians, because both its stage and bar had gathered the city's creatives and dissidents since its founding in 1964. The artistic director, executive director, and head of facilities approached London-based Haworth Tompkins, who had a two-decade track record in theatre design. The new Everyman could not just be an ordinary theatre: its architecture had to embody this heritage.

Design approach

Because of the tight inner-city site, the brief had to be distilled into a compact and adaptable design, and was developed from an understanding of the user's needs which included indoor air quality (IAQ) studies of the previous auditorium, and surveys of audience and staff members. Measurements were taken for 18 months in the auditorium of the old theatre prior to demolition. This included indoor and outdoor temperatures and relative humidity, and was used to establish the set-point temperatures for the new auditorium.

According to the architect, the goal of the design was to create a welcoming environment where the public would not feel that the technology and automation of the theatre caused a feeling of disconnection from the theatre experience.

If properly designed and operated, natural ventilation in the temperate climate of the UK can dramatically reduce electrical loads.[2] In the case of a

everyman

FIGURE 9.1.3
Ground floor plan.

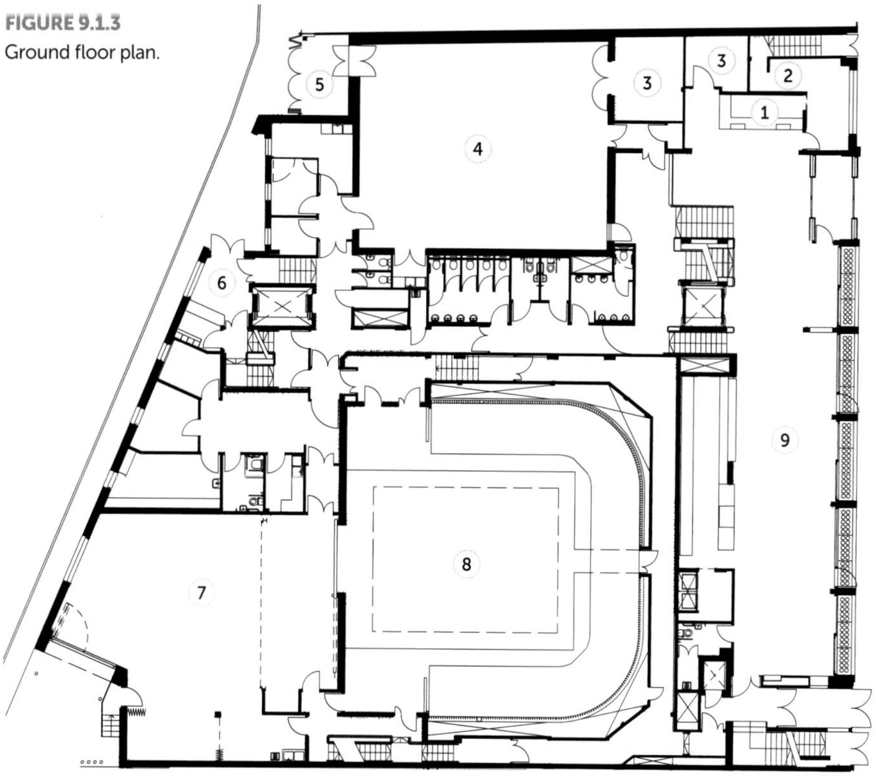

First floor plan.

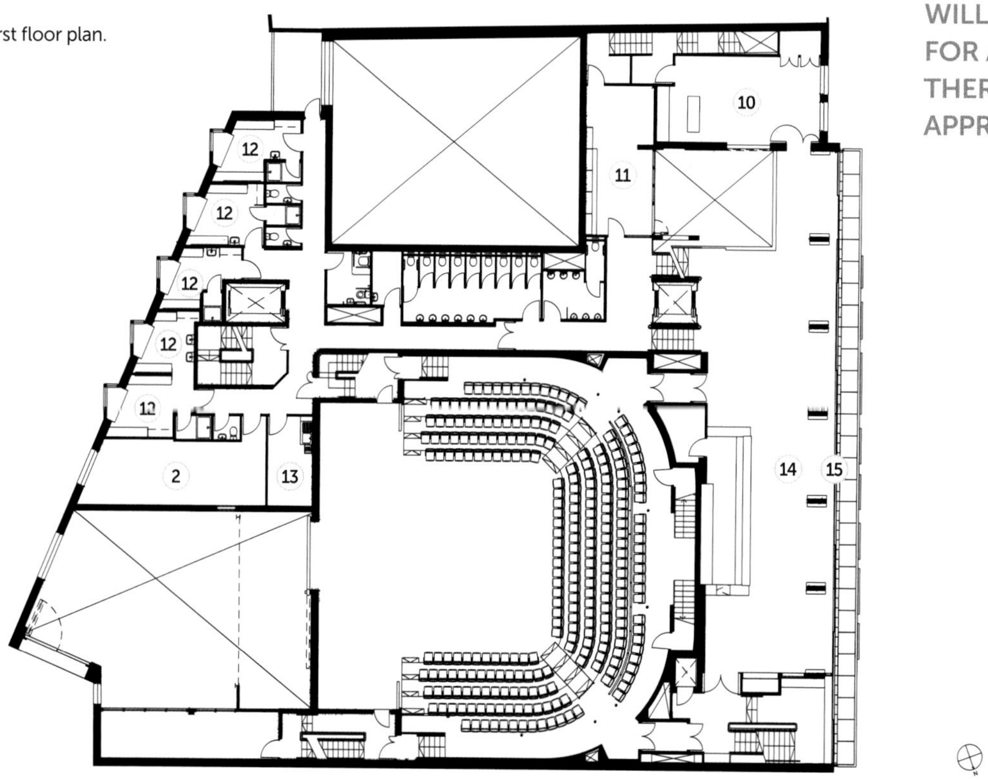

1 Box office
2 Office
3 Store
4 Youth and community studio
5 Refuse store
6 Stage door
7 Workshop
8 Stage
9 Cafe
10 Function Room
11 Writer's Room
12 Dressing Room
13 Office
14 Laundry
15 Bar
16 Movable metal sun screen

THE CLIENT WAS WILLING TO GO FOR AN ADAPTIVE THERMAL COMFORT APPROACH.

FIGURE 9.1.4
Rear elevation with intake vents for the plenum at street level.

theatre, natural ventilation has an added advantage, if attenuation is handled correctly, of being really quiet. The architect team also took on board the concerns of the actors that air-conditioned buildings negatively affected their performances and caused an off-putting background noise.

The Everyman presented a particular challenge, because its city centre site was compact and noisy. Space for a large plant room with chillers and air handling units (AHUs) was non-existent, which meant there was no alternative if the natural ventilation approach proved unworkable. But, 'you can't say whatever the weather, we will throw so much energy at it that it will rigidly hold to certain criteria', says architect Tompkins. With the client they discussed and agreed that with natural ventilation, seasonal comfort responses and expectations had to be much more aligned to the actual weather conditions, compared to a building with mechanical ventilation. The client was willing to go for this adaptive thermal comfort approach.

What this means in built form is that the theatre sits atop an approximately 1100-mm high air plenum. Air enters the plenum through long horizontal vents along the building's rear elevation, and is pulled through a labyrinth of concrete thermal mass to add coolth and attenuate sound, entering the 400-seat auditorium through vents below the seats. During a performance, the heat generated by actors, audience and lighting rises and then evacuates through an acoustically attenuated exhaust plenum integrated with the technical gallery level which connects to four large chimneys, forming distinctive external features.

Dampers adjust the volume of air intake according to the season. As a back-up, two AHUs fitted with air source heat pumps (ASHPs) are located within the plenum to top up heating or cooling with fan assistance if required. Filters slow down air movement, so no

FIGURE 9.1.5
West elevation with movable sunscreen etched with photographs of Liverpudlians.

filters were incorporated because there is minimal traffic at the rear of the building.

Below 21°C, fresh air supply is kept to a minimum and is triggered by carbon dioxide and temperature sensors. The heat pumps pre-heat auditorium air when the boilers are not running, but once the auditorium is full of people, no additional heat is required. If the auditorium reaches 24°C, the AHUs increase ventilation. Above 26°C, the system switches to mechanical cooling.

Strictly speaking, the Everyman is a mixed-mode building, because in winter the AHUs supply the ASHPs to heat the auditorium. In summer, the AHUs are not normally needed; adjustment of low- and high-level dampers suffices. The BMS monitors temperature and humidity and triggers the system when it hits its lowest set point.

The public art project on the facade, which is the Everyman's most widely published image, is in fact a sunscreen. Each panel can be internally adjusted to suit the needs of the interior space behind it. The shades in front of the offices tend to be changed more frequently because staff adjust them, while those at the theatre bar level are high up, and are rarely moved. The entrance foyers are vented through a large lightwell and opening screens on the west street elevation. Secondary spaces such as the smaller community performance spaces have their own mini-system, a separate street level intake and floor grilles, as well as vertical slot chimneys, while the rehearsal room has its own roof cowls as well as doors that open on to a terrace.

To meet the client's ambitious sustainability agenda, the architects and the engineers juggled competing priorities through an iterative process to achieve both the necessary voids for the plenums and shafts as well as sufficient thermal mass to make the system work. Collaboration over a shared vision was key,

FIGURE 9.1.6
Environmental section.

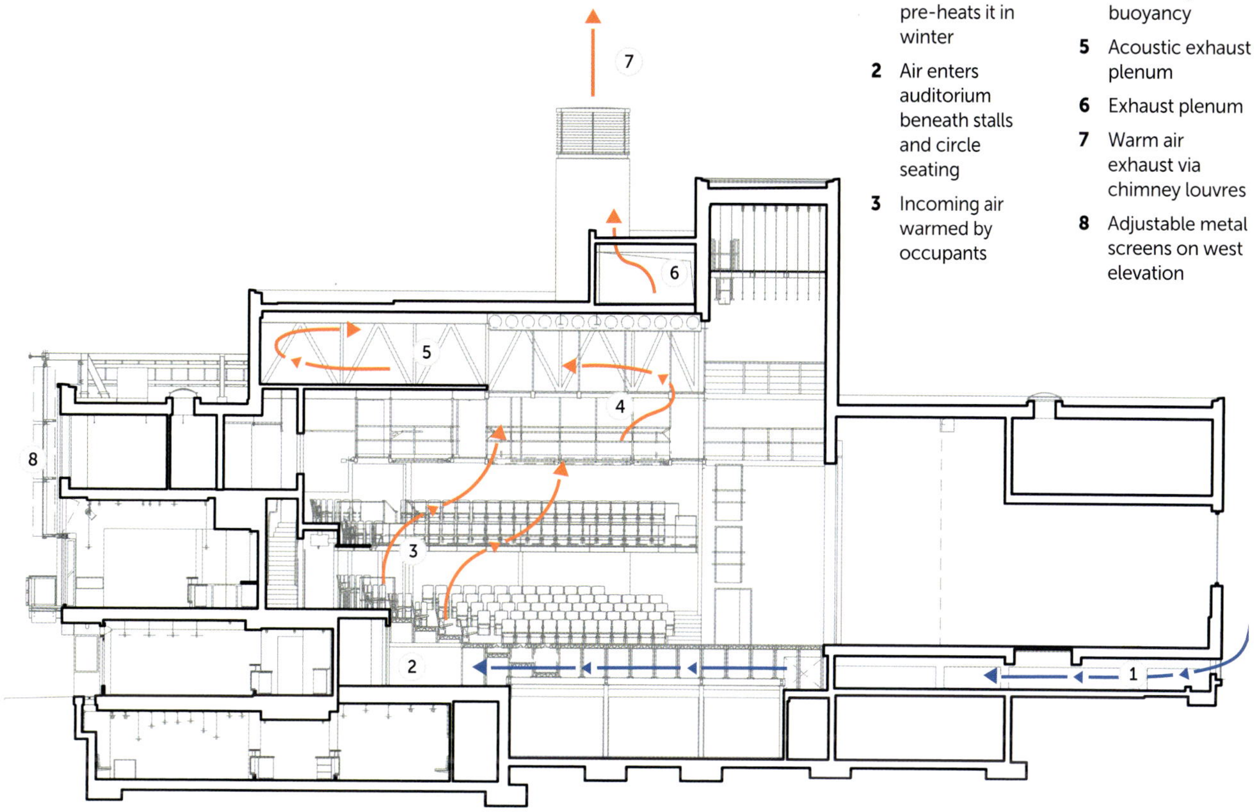

1 Thermal mass of plenum cools fresh air in summer and pre-heats it in winter
2 Air enters auditorium beneath stalls and circle seating
3 Incoming air warmed by occupants
4 Air further warmed by lighting, increasing its buoyancy
5 Acoustic exhaust plenum
6 Exhaust plenum
7 Warm air exhaust via chimney louvres
8 Adjustable metal screens on west elevation

and it clearly bore results. In addition to the ASHPs, other technical systems include gas-fired boilers and CHP, which initially worked effectively but developed technical complications in 2018. Because servicing from the manufacturer was patchy, at the time of writing it had yet to be returned to service.

Material reuse

Over 25,000 bricks were reclaimed from the former building and incorporated into the circular wall that encloses the auditorium. Salvageable timber and roofing materials were also reused. The services engineer estimated that the thermal mass of the reclaimed brick walls decreases temperatures in the auditorium by as much as 3°C.

Feedback and actual performance

'The building now works extremely well, even better than predicted,' says head of facilities Robert Longthorne, attributing this to the collaborative design process, an airtight, well-insulated envelope, and large thermal mass. 'The external envelope does the job because it creates a very stable internal environment,' he observes.

The handover process was critical. 'Commissioning takes so long; it is a learning process,' observes Longthorne, explaining that only in the second heating season did they get the BMS working properly, despite good support from most consultants and suppliers. They recall complaints of cold ankles during the first heating season. And, Longthorne notes, 'from year three the whole building has been working well. The heating and ventilation require minimal intervention. Most adjustments made in the early days were to do with tweaking fan speeds and trickling heat in to overcome drafts.' Longthorne acknowledges that the system must be actively managed in spring and autumn in response to variable weather conditions.

FIGURE 9.1.7
Circular walls of bricks reclaimed from the previous theatre on the site.

FIGURE. 9.1.8 EVERYMAN THEATRE ENERGY USE INTENSITY (EUI)

The total energy use intensity of the building matches the CIBSE Good Practice benchmark for theatres, based on figures published by the online CIBSE Benchmarking Tool. When looking at consumption by fuel, the building shows a reverse profile: heat consumption is well below the fossil energy use benchmark, while electricity consumption is higher. The theatre consumes significantly less energy than the CIBSE TM46 benchmark that a public building's DEC rating is based on.

EUI by fuel	Measured Y1	Measured Y2
Renewables	-15	-13
Heat	66	62
Electricity	105	100

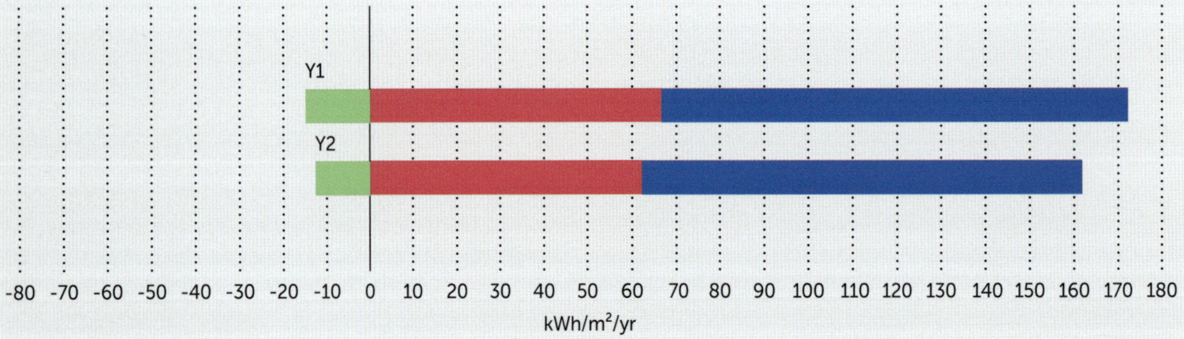

All post-occupancy energy and IAQ monitoring was undertaken by the client due to the commitment and initiative of the head of facilities; this was not part of the consultant's brief. As part of their Architects Declare commitment, the architects initiated monitoring and staff and audience surveys in 2019.

Energy monitoring was a key aspect of optimising the building's performance. However, Longthorne describes the challenge by saying, 'the biggest barrier to effective use of measuring and monitoring is the way the BMS saves and presents data'. Data is time-consuming to download and not easily manipulated into a format that can be easily interpreted and compared to industry benchmarks.

With over six years in operation at the time of writing, careful monitoring and active management of the system have demonstrated that in the UK's temperate climate, low-energy natural ventilation can work successfully, even in a building type with high internal heat loads from people and stage lights.

KEY MESSAGES

- **Pre-occupancy surveys of previous auditorium's audiences and indoor environmental quality (IEQ) data informed briefing of the new building.**
- **Committed strategic and detail-orientated head of facilities involved from brief-setting stage through to operation.**
- **Natural ventilation approach planned from outset so necessary shafts and chimneys were 'designed in' through an iterative process to accommodate volumes of air required.**
- **Fabric and thermal mass create stable internal environments.**
- **BMS data collection and interpretation time-consuming; ongoing adjustment of BMS by proactive FM during the first three years of operation with support of services engineer.**

KEY FACTS

25,000 BRICKS RECLAIMED FROM PREVIOUS THEATRE

ESTIMATED REDUCTION IN AUDITORIUM TEMPERATURE DUE TO THERMAL MASS OF BRICK WALL 3°C

MOVABLE METAL SUNSCREENS 105

50% LOWER HEATING CONSUMPTION THAN TYPICAL THEATRE AS PER CIBSE BENCHMARKING TOOL

THERMAL CHIMNEYS 4

CASE STUDY 2
MORELANDS ROOFTOP

TYPE NATURALLY VENTILATED ARCHITECTS' OFFICE

LOCATION ISLINGTON, LONDON

COMPLETION INITIAL REFURBISHMENT 1999; ROOFTOP 2013

CLIENT DERWENT

GROSS FLOOR AREA 1620 M^2

COST £4.4 MILLION

COST PER SQUARE METRE £2,716

NUMBER OF OCCUPANTS 230 (IN 2013)

SERVICES GAS BOILERS AND SOLAR THERMAL SUPPLY HEATING AND HOT WATER, AND A LANDLORD'S PHOTOVOLTAICS ARRAY CONTRIBUTES TO ELECTRICAL LOADS

AIRTIGHTNESS 5.0 $M^3/HR/M^2$ @ 50 PA

FORM OF CONTRACT TRADITIONAL JCT

ENVIRONMENTAL CERTIFICATION BREEAM OUTSTANDING

ARCHITECT AHMM

SERVICES ENGINEER PETER DEER & ASSOCIATES

STRUCTURAL ENGINEER AKERA ENGINEERS

CONTRACTOR CONTRAKT

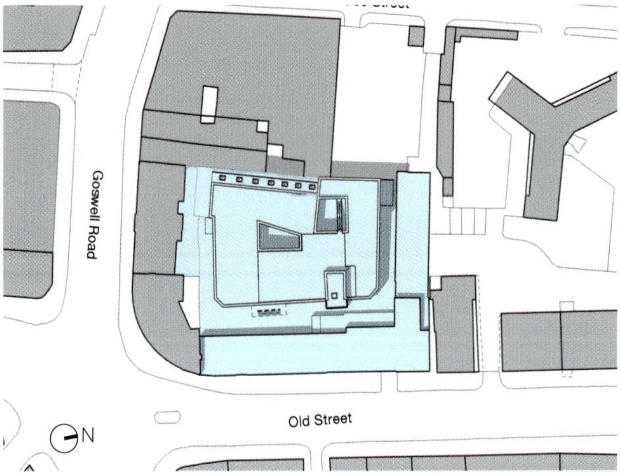

FIGURE 9.2.1
Location plan.

FIGURE 9.2.2: OPPOSITE
The new fifth floor features a 3.7m ceiling height with operable windows on actuators at high level to evacuate warm air via the stack effect.

An architectural practice's own office offers a testbed for design innovation. AHMM used its rooftop extension to Morelands, the London warehouse complex that the practice has occupied since 1995, to explore the viability of an adaptive comfort approach and its impact on people's wellbeing and productivity. Part retrofit and part newbuild, the BREEAM Outstanding project reconfigured existing offices on Morelands fourth floor and added a new fifth-floor penthouse with roof terrace to accommodate its growing staff of approximately 230 in 2013.

AHMM's staff salary costs represented, at the time, about 60% of the practice's operational costs, while utility costs were a tiny fraction of that (less than 1%). Therefore the primary driver was to create a quality work environment that would enhance productivity, while simultaneously demonstrating best practice in sustainability.

Design approach

Tall ceilings, good daylight without glare, operable windows and a lean approach to internal finishes and mechanical systems are central to AHMM's 1620 m^2 studio. This project exemplifies a nuanced approach to sustainable design that focuses on extending the life span of an existing building by enhancing environmental quality. It also includes a biodiverse brown roof, a solar thermal array and a PV array.

The office spans two floors with an internal lightwell: a reconfigured fourth floor of the existing brick and concrete building, and a new lightweight steel-frame fifth floor. The construction type was dictated by the structural limitations of the existing building. The existing fourth floor ceiling height of 2.9 m was increased to 3.7 m for the new fifth floor, and the exterior of both floors was rendered to increase insulation in the existing building and unify the design. Fourth-floor concrete soffits are exposed to exploit thermal mass.

BOLDLY DECIDING AGAINST MECHANICAL COOLING, AHMM ADOPTED A STRATEGY WHICH PUT STAFF IN CHARGE OF CONTROLLING THEIR OWN THERMAL COMFORT.

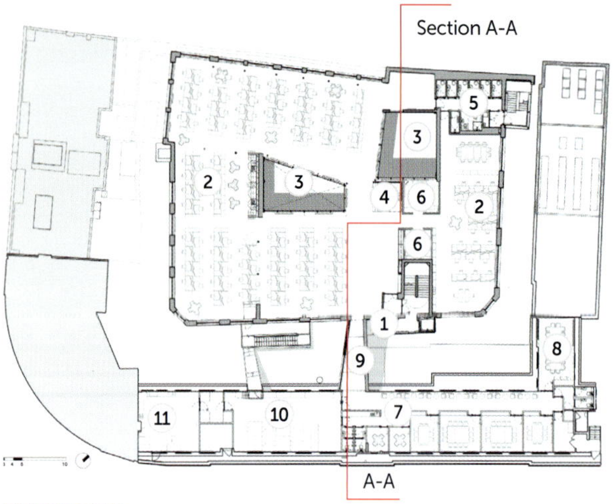

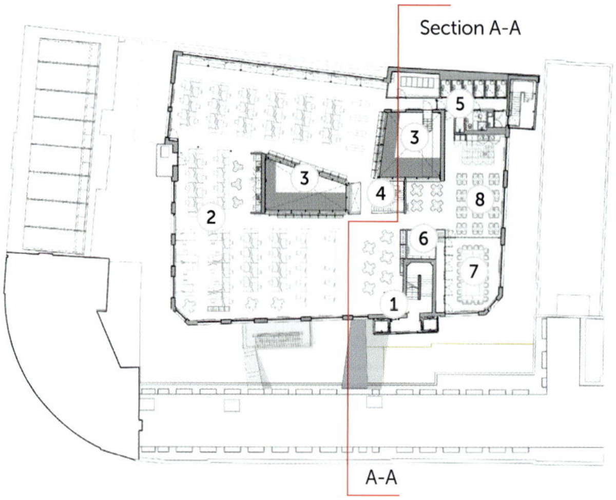

FIGURE 9.2.3

Fourth floor plan.

1 New lift
2 Open plan office
3 Lightwell
4 Internal staircase
5 WC
6 Printers
7 Meeting rooms
8 Project room
9 Bridge
10 Modelshop
11 Workshop

Fifth floor plan.

1 New lift
2 Open plan office
3 Lightwell
4 Internal staircase
5 WC
6 Printers
7 Meeting room
8 Refectory

Passive design informed the fenestration of the fifth-floor extension, with a reduced glazing ratio in the new top floor, compared to the existing floor below. The upper floor has better daylight because it benefits from sky views across the rooftops of nearby buildings, while the fourth floor windows face adjacent mansard roofs. Double-glazed curtain wall units are used throughout, and operable windows are positioned to enable cross-ventilation through the atrium.

Perimeter windows on both floors are comprised of an upper fixed pane with an operable top-hung outward opening unit at desk height. On the interior lightwell, all glazing is fixed except top-hung outward-opening units at soffit level, which enable natural ventilation. These are on actuators that are manually operated with a wall switch. All but the north elevation incorporate external blinds linked to UV sensors; a measure which has proven successful for controlling glare. The blinds lower automatically at certain sun levels. A manual override allows occupant control, and an anemometer triggers raising of the blinds in high winds. Staff are generally satisfied with the operation of the blinds, except on days which are both windy and sunny when the blinds are lifted.

Boldly deciding against mechanical cooling (with the exception of meeting rooms and the server room), AHMM adopted a strategy which put staff in charge of controlling their own thermal comfort. This approach assumed more relaxed comfort standards than BCO and CIBSE guidelines, anticipating that temperatures would fluctuate with external conditions within an acceptable range of adaptive comfort. As a result, mechanical plant was kept to a minimum: central heating and domestic hot water are delivered by gas boilers (one 60 kW boiler per floor) and solar thermal collectors. Morelands was also modelled at design stage for medium and extreme 2050 weather scenarios to understand potential future overheating risk.

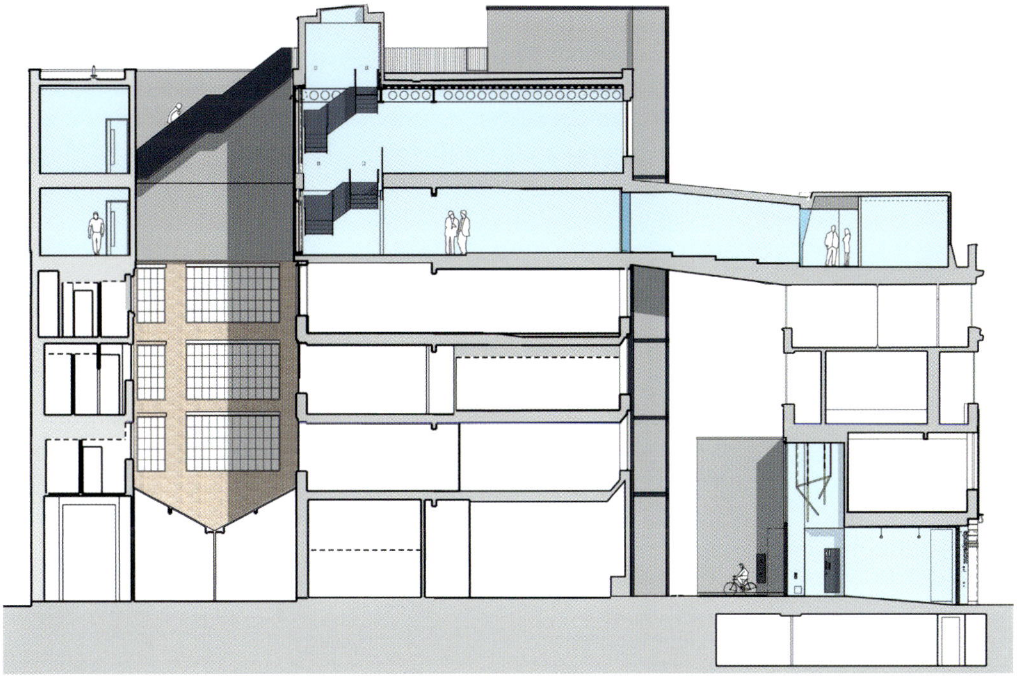

FIGURE 9.2.4
Section A-A.

FIGURE 9.2.5
Exterior courtyard with Morelands Rooftop on right.

FIGURE 9.2.7: OPPOSITE
View of the refurbished fourth floor with desk-height operable windows visible in the distance.

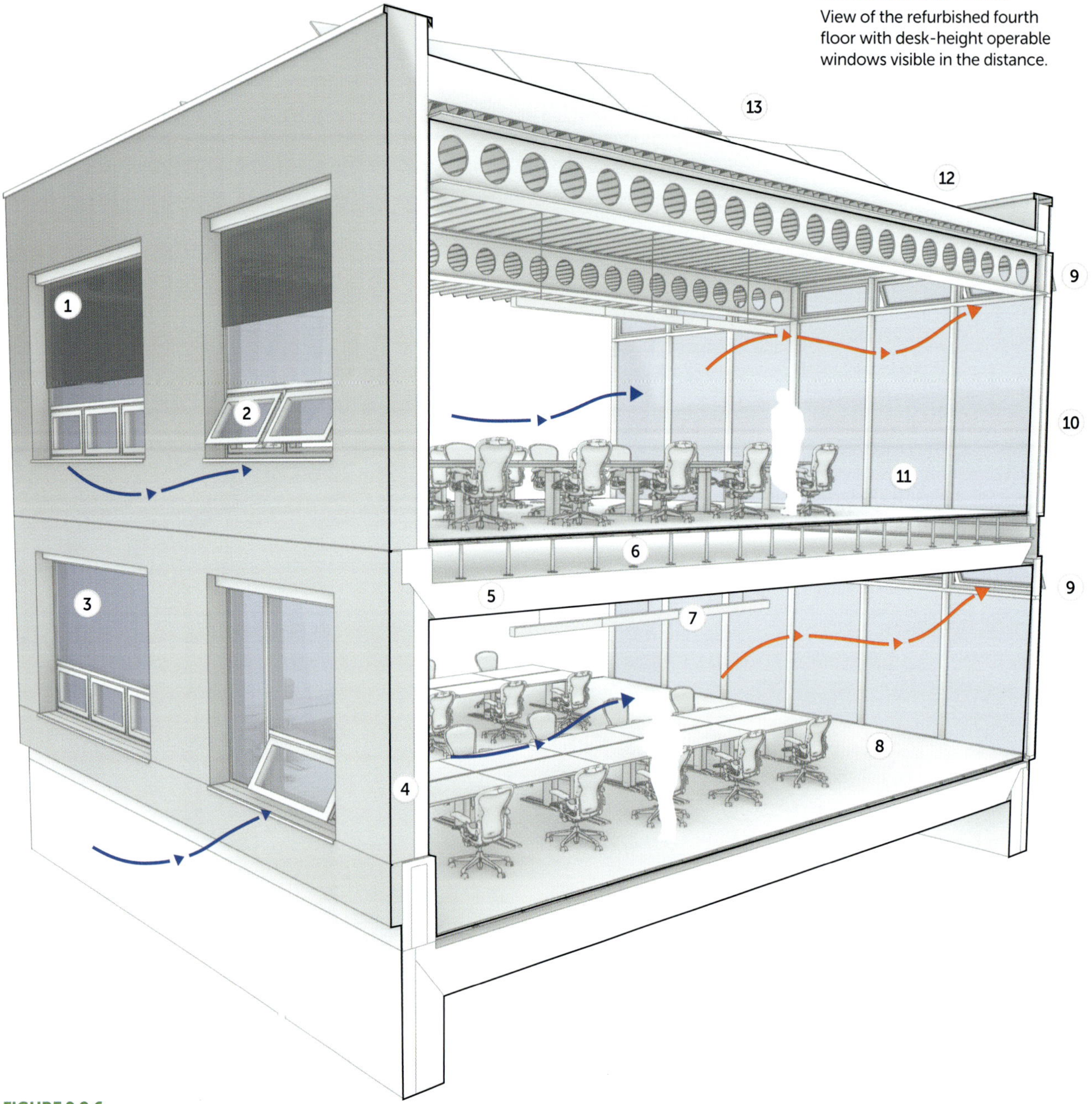

FIGURE 9.2.6
Environmental section.

1. **1** Automated blinds with manual override
2. **2** Manually operable window
3. **3** Fixed window pane
4. **4** EPS insulation and cement-free render
5. **5** Existing concrete slab (previously roof slab)
6. **6** Raised floor
7. **7** Lights with zoned controls and daylight dimming
8. **8** Existing 4th floor (ceiling height 2.9-m)
9. **9** High-level automated atrium vents
10. **10** Fixed glazing
11. **11** New 5th floor extension (ceiling height 3.7-m)
12. **12** Lightweight steel roof
13. **13** Rooftop PVs

Feedback and actual performance

A monitoring scheme was implemented approximately six months after occupancy, and this is ongoing. The practice sustainability group tracks temperature, CO_2 and relative humidity in six internal zones at 10-minute intervals, and readings are reported through the office intranet. Relative humidity and temperature of the external fabric are measured at three depths within the wall: 50 mm from the external face of the render, in the insulation, and on the external side of the internal finish. This makes it possible to understand heat flow and moisture movement in these monitored areas.

Thermal mass

The difference in construction type between the lightweight top floor and the heavyweight fabric of the existing building below offered an opportunity to directly compare the impact of thermal mass. When night cooling is used, the internal temperature drops about 3°C. Through modelling, the potential of night purging to exploit thermal mass was found to help alleviate overheating, with temperature peaks up to 2°C lower on the thermally massive fourth floor than on the lightweight upper floor.

Energy use

Post-refurbishment, primary energy consumption per desk space was reduced by almost 70%. At 173 kWh/m^2, the office is well below the CIBSE TM46 benchmark of 215 kWh/m^2, despite including the fifth-floor server room (which is not submetered). Actual measured consumption of the fourth floor (with no server room) is approximately one-third of the CIBSE benchmark, and heating energy consumption is well below the compliance calculation. Heating is off almost all year round, with the exception of between 7 am and 9 am on cold days.

In terms of benchmarking, it is difficult to compare the Morelands Rooftop energy consumption to RIBA Climate Change targets because the server room is not separately metered. This highlights the critical importance of submetering in order to accurately compare a project to industry benchmarks (see chapter 3). The large difference in energy consumption between the two floors suggests that if one were able to isolate the server room, the office would be much closer to the RIBA 2030 Climate Challenge targets for non-domestic buildings of less than 70 kWh/m²/yr.

Overheating

While the building performs very well in terms of energy use, overheating during the hottest summer days is a concern. Overheating results primarily from the high density of occupants and equipment. Data logging revealed that staff tended to open windows in response to rising temperatures. When indoor temperatures reached 25°C to 26°C, staff opened the windows no matter what, exacerbating overheating when outdoor temperatures were higher than indoors.

A study undertaken in summer 2015 revealed that the adaptive comfort summer temperature range was exceeded 3% of the time, well within the acceptable range. The office has a reserve of more than 40 desk fans that are used on the hottest summer days.

AHMM has identified several strategies which could reduce overheating. These include reduced technology such as replacement of PCs with remote servers and optimising natural ventilation through the use of sensors that indicate when windows should be opened.

The results of feedback were used to inform both staff and the FM team how to get the most out of the building. New posters instruct staff exactly when – and when not – to open windows. Sensor readings are also used by FM staff to determine the extent of night purging on a given day.

Occupant surveys

Building Use Survey (BUS) – conducted both before the move and 14 months after refurbishment – reveal

FIGURE. 9.2.8 MORELANDS ROOFTOP ENERGY USE INTENSITY (EUI)

Morelands Rooftop offices data reveals significantly higher energy consumption than the other office buildings featured, because its data is skewed by an energy-hungry server room which is not separately submetered. This highlights the critical importance of submetering in order to compare a project to industry benchmarks. If the server room were decoupled from total consumption, Morelands Rooftop's energy use would most likely be only slightly higher than the Keynsham Civic Centre EUI and is approaching the RIBA 2030 Climate Challenge benchmark.

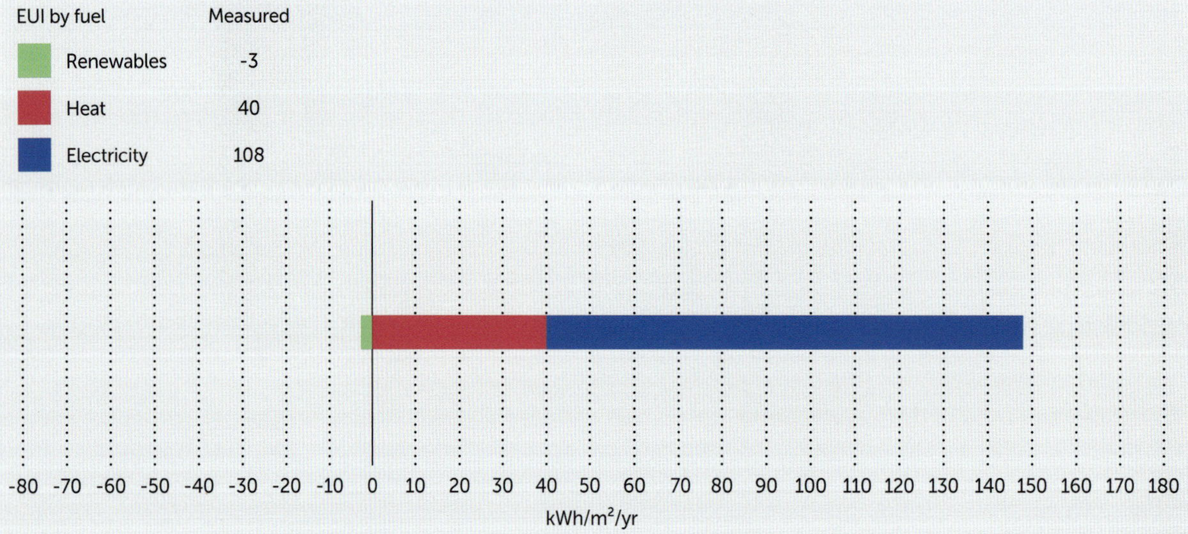

a high degree of satisfaction from staff despite some 'overheating' and high CO_2 levels in summer. Temperature and air quality in summer were the only attributes that occupants deemed unsatisfactory, yet surprisingly, the BUS survey found that staff rated the lightweight fifth-floor summertime temperature and air quality better than the fourth floor, despite higher temperatures on the upper floor. The research team attributed this to the 'forgiveness factor' of the taller ceiling heights, increased daylight and broader vistas from the top floor.

Monitoring studies have validated AHMM's design approach: staff surveys confirm that good architecture with adaptive comfort is more pleasant to inhabit, and a major benefit of this approach is that it enables building users to take responsibility for their own comfort, with the additional benefit of more satisfied occupants.

Furthermore, AHMM concluded that well-managed naturally ventilated buildings in central London can provide productive work environments. AHMM's Morelands Rooftop markedly exceeds current comfort thresholds. Based on the evidence gathered within its own offices, AHMM has been able to confidently promote an adaptive comfort approach in subsequent projects.

KEY MESSAGES

- **Well-designed building conFIGUREuration, materiality and daylighting enhance the comfort perception of occupants, reducing the need for energy consuming technical systems.**
- **Thermal mass can help balance out high temperature peaks (thermally massive fourth floor compared to lightweight fifth floor).**
- **Occupants tend to open windows when indoor temperature reaches 25°C to 26°C even when outdoor temperatures are higher, exacerbating overheating; clear instructions or indicators on when to open windows are essential.**
- **BUS survey confirmed that generous volume, daylight and views enhanced staff satisfaction and increased tolerance of summertime air quality and overheating.**
- **Feedback – both quantitative and qualitative – provides designers with knowledge and evidence about how buildings work.**

KEY FACTS

70% REDUCTION IN PRIMARY ENERGY CONSUMPTION PER DESK SPACE AFTER REFURBISHMENT

3° DAYTIME PEAK TEMPERATURE DROP FROM PEAK DUE TO NIGHT PURGING

NUMBER OF ZONES WITH IAQ SENSORS **6**

<1% OF OPERATING BUDGET EQUAL TO ENERGY COSTS

4TH FLOOR ACTUAL ENERGY CONSUMPTION COMPARED TO CIBSE GUIDE F TYPICAL BENCHMARK **1/3**

CASE STUDY 3
CARROWBRECK MEADOW

TYPE PASSIVHAUS HOUSING

LOCATION HELLESDON, BROADLAND DISTRICT, NORFOLK

COMPLETION 2016

CLIENT BROADLAND GROWTH LIMITED

GROSS FLOOR AREA 14 HOMES, RANGING FROM 70 M^2 TO 130 M^2

COST PER SQUARE METRE £1,400 (THREE-STOREY HOUSE); £1,750 (HOUSE WITH FIRST-FLOOR TERRACE)

SERVICES HEATING AND HOT WATER SUPPLIED BY GAS BOILER; MHRV; ONE RADIATOR/FLOOR AND HEATED TOWEL RAIL (TWO RADIATORS IN GROUND FLOOR IN ONE HOUSE TYPE)

AIRTIGHTNESS 0.3–0.5 M^3/HR/M^2 @ 50 PA ACROSS 14 HOMES

FORM OF CONTRACT TWO-STAGE DESIGN AND BUILD

ENVIRONMENTAL CERTIFICATION PASSIVHAUS

ARCHITECT HAMSON BARRON SMITH

SERVICES ENGINEER NPS

STRUCTURAL ENGINEER ROBSON LITTLE

LANDSCAPE ARCHITECT HAMSON BARRON SMITH

CONTRACTOR RG CARTER

FIGURE 9.3.1
Site plan.

FIGURE 9.3.2: OPPOSITE
The houses were sited to enable retention of mature trees and are all orientated within 15 degrees of due south so that PVs would be optimally positioned if added in the future.

Carrowbreck Meadow is a 14-unit RIBA-award-winning Passivhaus development approximately three miles northwest of Norwich city centre. A thoughtful site plan retains mature existing trees and groups the houses in threes, making the outdoor spaces around and between them add up to more than the sum of the parts, and at the same time disguising the high three-car-per-unit parking ratio. In terms of design, a combination of black timber cladding and white lime render with a steep 50-degree roof pitch references historic barns seen throughout Norfolk, according to Hamson Barron Smith associate director Sarah Lewis. The units are nestled among trees that provide a sense of intimacy and enclosure.

Carrowbreck Meadow is luxury housing with a mix of two-, three- and four-bedroom homes ranging in size from just over 60 m^2 to almost twice that, at 118 m^2. Just under half of the units are designated shared equity. Numerous aspects of this development are easily replicable and point the way to housing of the future. The homes feature well-integrated timber brise-soleils, external Venetian blinds and rainwater butts.

Carrowbreck Meadow was the first of several local authority-built housing schemes to complete in Norfolk. At the time, Norwich City Council had a large Green Party membership advocating for low-energy homes, and Passivhaus offered a proven performance record. 'Based on experience, I was able to demonstrate the measured performance of Passivhaus homes,' explains Lewis.

Design approach

Solar orientation and retention of mature trees determined the site layout, and all units are located within 15 degrees of due south. For a certified-Passivhaus development, what is immediately striking is the small amount of south-facing glazing. About

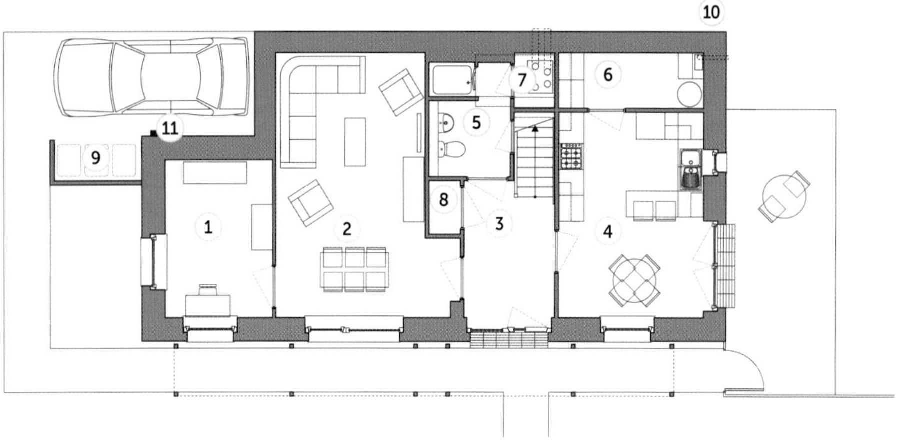

FIGURE 9.3.3
House plans – Typical ground floor plan.

1 Study
2 Living/ dining
3 Hall
4 Kitchen/ breakfast room
5 WC/ shower
6 Utility
7 MVHR cupboard
8 Cloaks
9 Bin/ recycling store
10 Rainwater tank
11 Electric car charging point

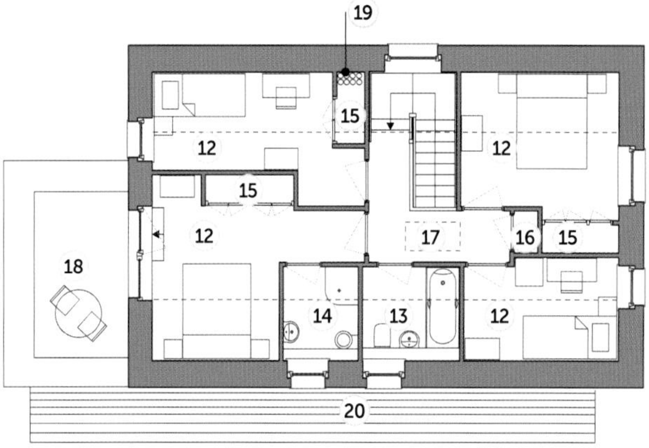

House plans – Typical first floor plan.
12 Bedroom
13 Bathroom
14 Ensuite
15 Wardrobe
16 Cupbaord
17 Loft access
18 Terrace
19 MVHR ducts
20 Brise-soleil

40% of the glazing is on the south facade, largely shaded by fixed brise-soleils. Only 2% faces north, and just under a third is orientated east or west. East- and west-facing windows incorporate external Venetian blinds, and the stairwells of all units have a tilt-and-turn window to facilitate stack ventilation. All windows are triple-glazed, opening inwards to allow for optimal detailing of external installation over the frames. Windows can be left open with external blinds lowered for solar shading. The blinds are electrically operated via a manual switch.

'The houses must be very robust because they were not designed for occupants who are super-into Passivhaus,' explains Lewis. Feedback from Lewis's prior experience informed the design of these homes: 'The owner often used the blinds opposite to the way they were intended: down in winter for privacy because the trees were bare and up in summer to enjoy the sunshine.'

At Carrowbreck, the limit for overheating was modelled to allow an internal temperature of greater than 25°C only 1% of the time, well below the 10% limit permitted for Passivhaus certification. Passive design features such as limiting south-facing glazing, deep window reveals (228 mm minimum) and timber pergolas carefully calculated as brise-soleils using sun-path analysis all help to control solar gain.

For the PHPP calculations used to design to Passivhaus standard, the shading impact of the trees was calculated on a window-by-window basis, taking into account the height and spread of any trees contributing shading throughout the seasons.

Simplicity was the driving principle behind the detail design at Carrowbreck. External walls are load-bearing hollow clay blocks, and insulation wraps under the concrete slab and over the loft spaces to prevent thermal bridging. Both exterior render and interior render are lime-based, allowing the walls to breathe.

FIGURE 9.3.4
Deep window reveals, external blinds and brise-soleils control solar gain.

FIGURE 9.3.5
Motorised external blinds, controlled manually via a switch by residents, limit solar gain.

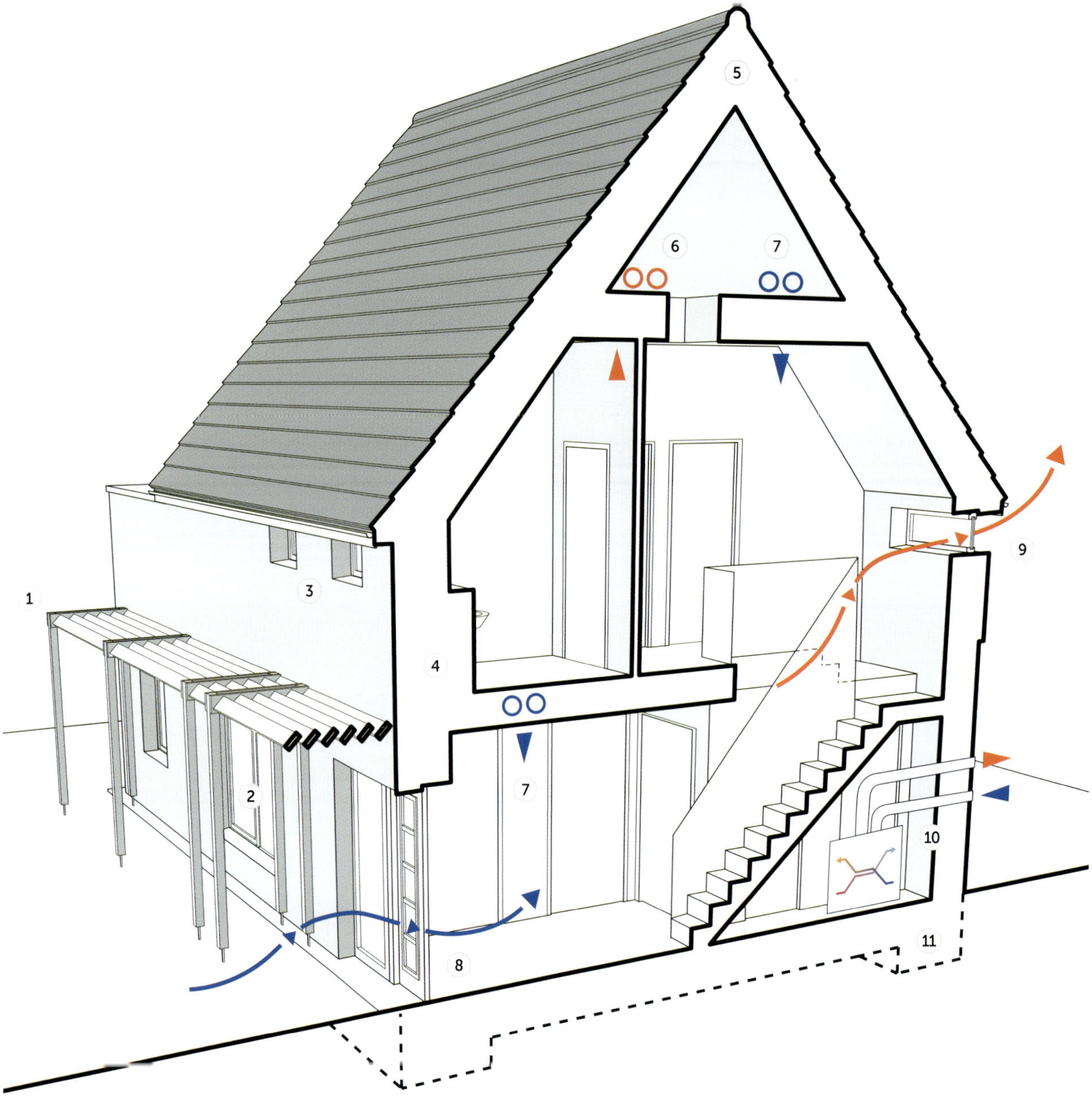

FIGURE 9.3.6

Environmental section.

1. Timber pergola on south facade
2. Electrically-operated blinds with manual switch on east and west facades
3. Deep window reveal
4. Load-bearing hollow clay blocks insulated externally with lime-based render inside and out
5. Thermal bridge-free ridge detail
6. Extract air vents
7. Fresh air vents
8. Low level summertime air intake
9. High level summertime air outlet
10. Heat recovery ventilation unit in downstairs bathroom
11. Thermal bridge-free wall and foundation detail

WHAT IS NEEDED IS TO DESIGN HOMES THAT PEOPLE CAN USE AS INTENDED WITHOUT HAVING TO THINK ABOUT 'GREENING' THEIR BEHAVIOUR.

FIGURE 9.3.7

Summer overheating graph. During a heatwave in summer 2017, stable temperatures were maintained inside when outdoor temperature exceeded 33°C.

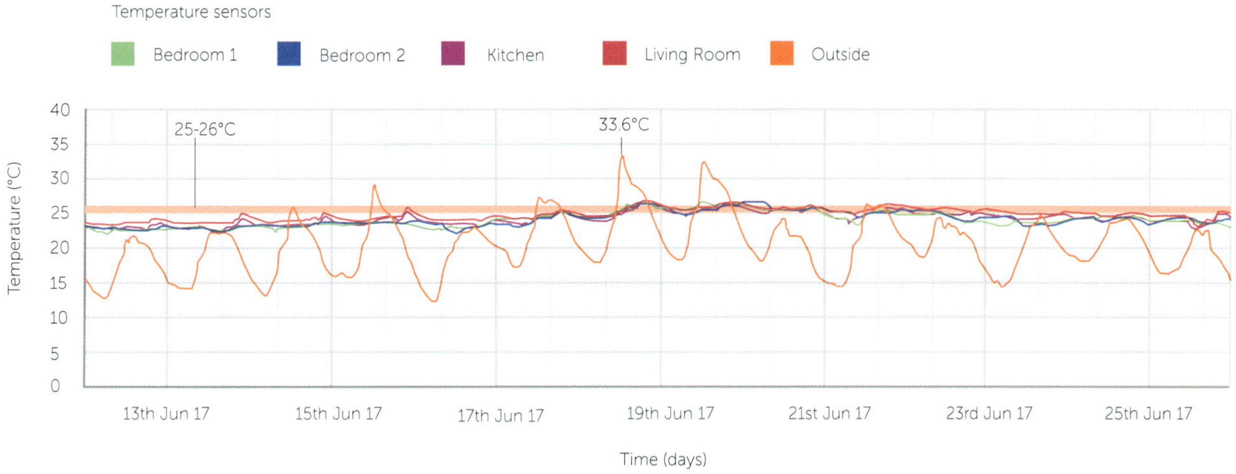

Mechanical ventilation and heat recovery (MVHR) units, used year-round with summer bypass, are typically located in the ground-floor bathroom with full access to the unit, and the filter is positioned at a convenient height so it is easy to replace. Third-party MVHR commissioning was a contractual requirement.

Client commitment to Passivhaus was critical to the success of the scheme, and Passivhaus certification was made a planning condition to ensure that it would not be lost over the course of the project. Carrowbreck Meadow was delivered through a two-stage design and build contract, and the project was the contractor's first experience with Passivhaus.

The architects worked closely with the contractor to develop a wall element which maintained the U-value required for Passivhaus. For the timber walls, the performance of the insulation was increased to compensate for the thinner insulation zone and to account for the thermal bridges created by the fixings for the timber battens. This was fed back into PHPP as part of an iterative technical design process with the contractor to develop the most efficient solution in terms of performance, buildability and cost.

A key principle throughout the project was to minimise automation and complex user controls in the homes. 'People are allergic to automation in residential environments,' says Lewis. 'Lots of unfamiliar user controls which require people to change their habits just don't work. In Britain, people are not accustomed to using shutters and blinds to control solar gain, so often use them incorrectly,' she says. What is needed is to design homes that people can use as intended, without having to think about 'greening' their behaviour.

Construction quality

Lewis notes that 'complete buy-in' from the contractor facilitated the construction process. For example, the entire project management team completed a four-day certified Passivhaus Tradesperson course, and the contractor appointed both a designated Passivhaus champion and a separate airtightness champion. The site foreman organised trade-specific talks targeted at critical areas, such as detailing around door and window openings, and not puncturing the airtightness layer during first and second fix. This was clearly successful, because every unit passed its airtightness test the first time.

Blown insulation was installed in the roof cassettes on site, and correct installation was verified with thermography.

Feedback and actual performance

At Carrowbreck Meadow, the feedback process was entirely led by the architects, and included semi-structured interviews with residents, collection of utility bills, remote web-based monitoring and web-based occupant surveys with a 65% response rate. Lewis developed the survey based on the BUS Methodology. Monitoring of internal environmental conditions was undertaken on two of Carrowbreck's affordable homes, and energy bills were collected for all units. A quarterly newsletter updates residents with relevant news and maintenance tips.

In terms of energy, gas for space heating and hot water use, and electricity use are monitored. Indoor environment quality (IEQ) parameters being monitored include CO_2, temperature and relative humidity. All measurements have fallen within the best practice range, with the exception of minimal overheating during the 2017 summer heatwave, still well within the Passivhaus threshold of not exceeding 25°C for more than 10% of the year. According to one resident: 'Air quality is amazing ... We all have wonderful sleeps at night which we believe is due to air quality. The consistent temperature of the house is perfect.'

Two-thirds of residents participated in a survey undertaken during the second year of occupancy. Almost 90% of respondents were satisfied with the design and layout of the development and their home. A surprising 85% said they would now prefer to live in a

FIGURE. 9.3.8 CARROWBRECK MEADOW ENERGY USE INTENSITY (EUI)

The Carrowbreck Meadow houses achieved a 16% lower heat demand in use than predicted, while electricity use is an average taken between the highest and lowest monitored households. This development almost meets the RIBA 2030 Climate Challenge targets. The client opted to invest in a robust fabric and future proofed for renewables. The homes are wired with a conduit at roof level so that PVs can be added by the owners. Where renewables are installed, operational carbon emissions could easily reach net zero performance.

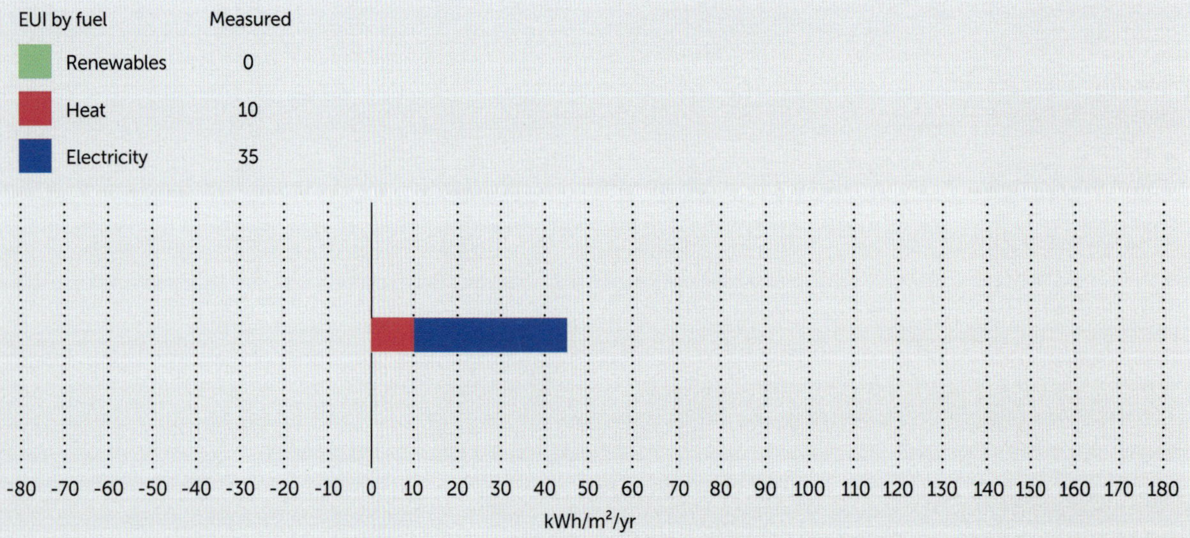

Passivhaus or would only move to another Passivhaus, rating their home as comfortable in both summer and winter. A similar percentage found the heating as easy or easier to use than systems they were familiar with. As for the MVHR, just over half of residents rated it at least 'quite easy' to use, and 85% ranked changing MVHR filters as 'quite' or 'very easy.' All respondents were satisfied or very satisfied with their energy bills.

The fact that Carrowbreck Meadows is a small development with just 14 units suggests that it would be relatively straightforward to monitor and gather energy and internal comfort data as well as feedback from residents. However, problems with the web-based monitoring system and data dropouts meant that the process was not straightforward on a site-wide basis, and as it was not a contractual requirement, the issue was not solved. The architects remain proactively involved in the post-occupancy evaluation of these homes and are in regular contact with residents to gather annual energy bills and provide informal opportunities for further feedback on the comfort of the homes. Because the technology has improved substantially in the last five years, it is now much easier to plan ahead for monitoring, according to Lewis.

The variety of feedback processes used at Carrowbreck Meadow provided follow-through and direct engagement with residents, which enabled troubleshooting of problems and an on-site presence as residents became familiar with their new homes. It is challenging for residents to digest user manuals when first moving in, and this information is better communicated once people are settled.

KEY MESSAGES

- **One year of post-occupancy energy and IEQ monitoring required for Passivhaus certification ensured that feedback was built into project planning and validation delivered predicted performance.**
- **Contractor upskilling of all trades through Passivhaus workshops enhanced awareness and construction quality.**
- **Thermography used to verify continuity of insulation.**
- **Contractual requirement for third-party commissioning of MVHR provided quality control.**
- **Occupant surveys undertaken at architects' initiative and expense.**
- **Site-wide data collection plagued by problems with the remote monitoring platform, but technology has subsequently improved substantially.**

KEY FACTS

16% SPACE HEATING DEMAND BELOW DESIGN TARGET DURING FIRST MONITORED WINTER

RESIDENTS SURVEYED WHO WOULD PREFER TO LIVE IN A PASSIVHAUS IN FUTURE 85%

NUMBER OF MATURE TREES RETAINED ON SITE 122+

30% RETURN ON GROSS DEVELOPMENT VALUE TO PUBLIC SECTOR

AMOUNT OF TIME INTERNAL TEMPERATURE MODELLED TO SURPASS 25° 1%

CASE STUDY 4
ROCKY MOUNTAIN INSTITUTE INNOVATION CENTER

TYPE NATURALLY VENTILATED OFFICE

LOCATION BASALT, COLORADO

COMPLETION 2016

CLIENT ROCKY MOUNTAIN INSTITUTE

GROSS FLOOR AREA 1,450 M^2

COST £10.4 MILLION ($13.6 MILLION)

COST PER SQUARE METRE £7,183

NUMBER OF OCCUPANTS 35 STAFF + UP TO 80 FOR CONFERENCES

SERVICES IN-FLOOR ELECTRIC RADIANT HEATING, AIR-TO-AIR HEAT EXCHANGER, INDIVIDUAL HYPERCHAIRS, CEILING FANS (NO COOLING)

AIRTIGHTNESS 0.36 ACH^{-1}

FORM OF CONTRACT INTEGRATED PROJECT DELIVERY

ENVIRONMENTAL CERTIFICATION LEED PLATINUM, PASSIVHAUS, LIVING BUILDING CHALLENGE (PETAL CERTIFIED, VERSION 2.1)

ARCHITECT ZGF WITH GRAYBEAL ARCHITECTS

HIGH-PERFORMANCE DESIGN CONSULTANT ARCHITECTURAL APPLICATIONS

MEP + IT ENGINEER PAE CONSULTING ENGINEERS

STRUCTURAL ENGINEER KPFF CONSULTING ENGINEERS

LANDSCAPE ARCHITECT DHM DESIGN

LIGHTING DAVID NELSON & ASSOCIATES

CONTRACTOR JE DUNN

COMMISSIONING RESOURCE ENGINEERING GROUP

The Rocky Mountain Institute (RMI) headquarters in Colorado illustrates what can be achieved when an ambitious client tackles zero carbon building design with a holistic approach, and monitors performance to ensure that the building delivers on its aspirations.

Zero carbon buildings are 'challenging to deliver', particularly in Colorado's extreme alpine climate. However, this LEED Platinum Passivhaus-certified building is zero energy for both regulated and unregulated energy use, achieved by integrated design from project inception through to completion, including a rigorous performance contract.

The Rocky Mountain Institute is an environmental NGO whose mission is to promote market-based solutions that shift the global economy off fossil fuels and towards a low carbon future. Headquartered in Colorado, RMI employs more than 200 people in five offices, from Washington, DC to Beijing.

FIGURE 9.4.1
Site plan.

FIGURE 9.4.2: OPPOSITE
South-facing glazing in the work areas if fitted with blinds that automatically adjust to to six different pre-set positions.

THE END OF GROWTH
GREEN TO GOLD
The Innovator's Dilemma
A GREAT ARIDNESS
PRIZE

FIGURE 9.4.3

The northern elevation (shown) that fronts on to the car park has an 18% glazing ratio, compared to 52% on the south side facing the river.

FIGURE 9.4.4

Ground floor plan with north-facing entrance and large glazed facade to the south, overlooking the river.

1 Lobby
2 Break out room
3 Conference room
4 Dining
5 Media room
6 Mechanical
7 Open Office
8 Bike room
9 Office
10 Open Office
11 Conference room
12 Reading room
13 Terrace

First floor plan features a south-facing open-plan office with spectacular views into the mountains.

FIGURE 9.4.5
South-facing elevation features a light shelf with phase-change material and external metal blinds to control heat gain.

With its new Colorado headquarters, RMI was committed to a design approach that would be replicable, and that the wider industry could learn from. By 2035, approximately 75% of office buildings in the US will be new or renovated, so many aspects of the project's approach could be replicated.

RMI wanted to move the industry forward by sharing its process, in particular the use of performance contracting. Durability was another key driver, and the building has targeted a 100-year design life, a considerable improvement over the 60-year horizon that represents current best practice in the UK, and is rarely achieved in reality.

Design approach

At the RMI Innovation Center, ZGF Architects have pushed the limits of passive design to create congenial office and meeting spaces. Orientated south to maximise solar gain, the linear two-storey CLT structure is clad in local sandstone, untreated juniper and zinc.

The 16-m deep building opens to the south and benefits from excellent daylight, and cross-ventilation. The highly insulated airtight external envelope incorporates quadruple-glazed windows. The glazing ratio on the south facade is 52% compared to 18% to the north.

Sun-responsive external blinds are BMS-controlled to six different fixed positions to manage heat gain and glare. Concrete floors provide thermal mass and are linked to the BMS so that when the slabs reach 25°C, the blinds automatically lower and, if needed, windows on actuators open. Windows and blinds have manual overrides, and windows are equipped with a green light that indicates when they can be opened. During summer, windows remain open for night cooling until the temperature of the slab drops to 13°C.

In winter, a small system of underfloor radiant electric heating tops up the internal temperature on

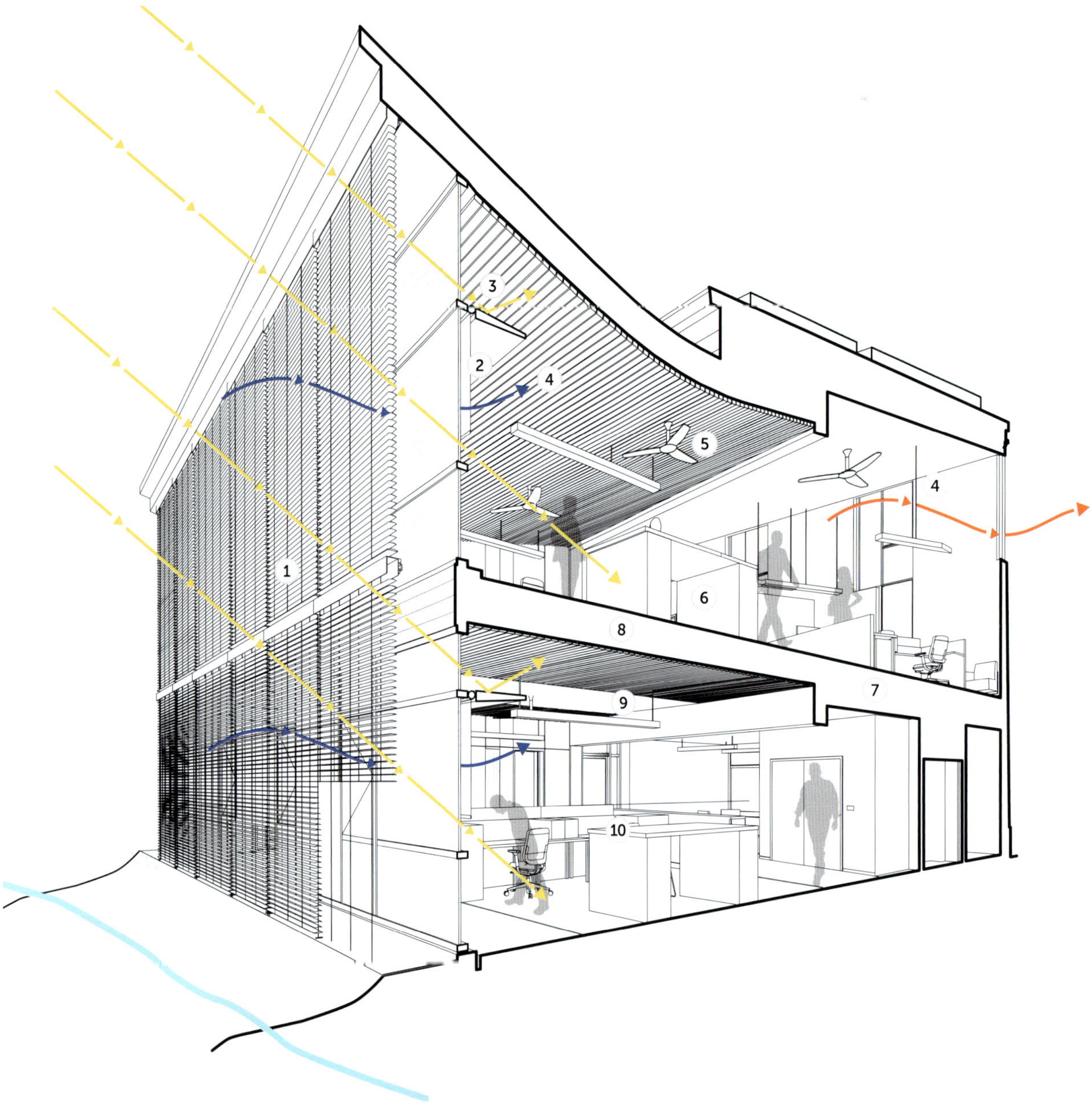

FIGURE 9.4.6

Environmental section.

1 Automated exterior blinds
2 Interior fabric blinds
3 Light shelves
4 Combination of cross and stack ventilation with actuated windows on south and north facades
5 Ceiling fans
6 Shallow 15-m floor plate
7 Supply from heat recovery ventilation
8 Supply from background mechanical ventilation
9 Efficient ambient LED lighting
10 Individual LED task lights

PHOTOVOLTAIC PANELS ON THE ROOF PRODUCE TREBLE THE REQUIREMENTS OF THIS ALL-ELECTRIC BUILDING, AND ENOUGH ENERGY TO POWER SIX CARS.

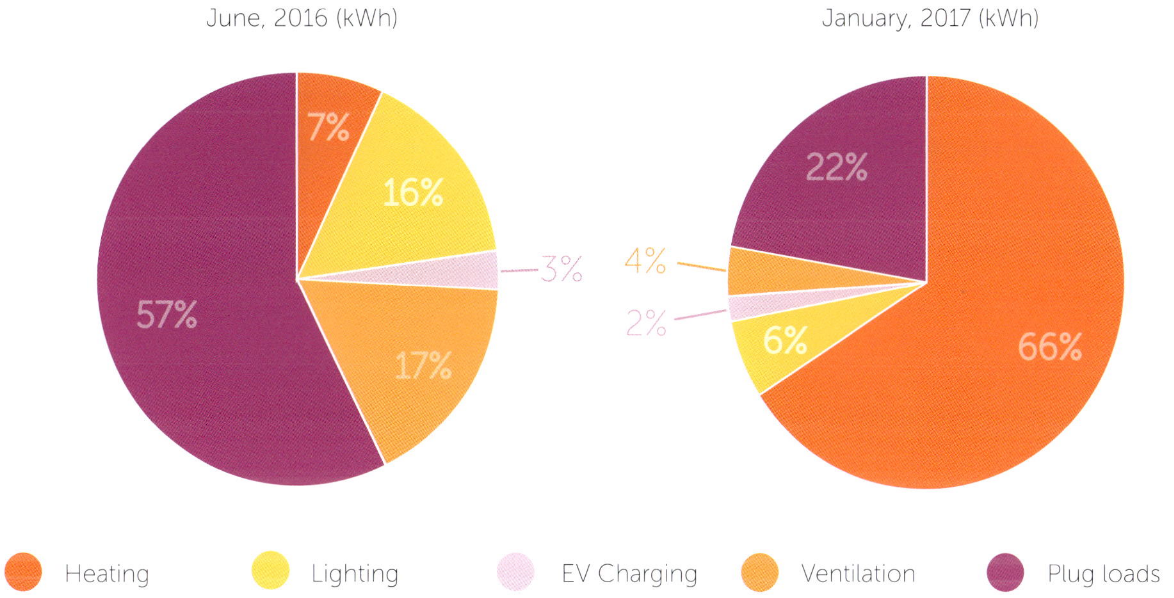

FIGURE 9.4.7
Energy end-use breakdown. Energy uses vary dramatically between the seasons due to increased heating and plug loads in winter.

the coldest days, supplying heat directly to workspaces rather than conditioning the entire internal volume. If this is insufficient, battery-operated desk chairs by HyperChair are fitted with heaters and fans, and staff can control both heating and overhead fans manually or from their smartphones. Two air-to-air exchangers provide fresh air based on CO_2 sensors, and ceiling fans facilitate air movement on hot summer days.

Light shelves run the length of the south elevation and bounce daylight deep into the plan; they also contain phase-change materials to increase thermal mass. Artificial lighting is supplied through ceiling-mounted LED fittings, supplemented by individual task lighting at every desk.

For an adaptive approach to thermal comfort, RMI opted to allow wider temperature swings than a conventional office. Most systems in America are set to a narrow comfort band of 24°C to 26°C, while at the RMI Innovation Center temperatures can drop as low as 18°C and climb up to 28°C on exceptionally hot summer days.

Photovoltaic panels on the roof produce treble the requirements of this all-electric building, and enough energy to power six cars.

The building was constructed using integrated project delivery (IPD), an incentive compensation contract whereby all consultants, contractor and subcontractors pay into a reserve at the beginning of project. If the project is completed on or below budget, the reserve is distributed back, with 50% to the client and 50% to contractor and consultant team.

Cost

RMI calculated the uplift in cost to meet net zero carbon standard at approximately 10.8% of the contract value, or £646 per m² ($86 per sq. ft) and estimated that the payback in savings would take less than five years. In fact, it took 3.9 years. Savings result from reduced energy and maintenance costs, and increased staff productivity and satisfaction, which RMI calculated at 3%, based on over 30 studies.[3]

According to RMI's year one report, the soft cost premium for delivering the zero carbon building was £255 per m² ($32 per sq. ft), due primarily to four factors: the integrated design process, which involved contractor and subcontractor costs pre-construction; the use of specialist consultants in high-performance design, additional time for trialling IPD, and a learning curve for Revit and IESVE modelling. The design team's view is that this cost premium will be 'eroded by market forces as scaling occurs'.

Feedback and actual performance

Because energy and internal comfort monitoring was integral to the project from the outset, the building was fitted with over 120 sensors. Every desk is submetered to manage plug loads. Energy data from the first year of occupancy showed that the building was operating at below predicted levels. In summer, plug loads account for more than half of the building's energy loads, while in winter electric resistance heating is the primary load. A battery storage system powered by the PV array helps to maintain peak demand under 50 kW.[4]

The small RMI team had extensive pre-occupancy training, regular meetings to gather feedback after the move, and a 'green team' to address real-time issues. Workspaces are programmed for automatic shutdown from 7 pm to 7 am, with individual manual overrides.

As feedback has shown on other projects, the biggest challenge was integrating the building systems with its automated controls. Customisation of software by the engineers and the vendor was required to calibrate the settings and ensure that blinds would not be damaged by a sudden gust of wind. Sensors on the window actuators also had to be adjusted, because they drew electricity when not in use. It was also noted that control systems needed to be simplified.

A robust commissioning process was critical to the success of this project in achieving performance targets. The client hired a commissioning agent for the

FIGURE. 9.4.8 RMI ENERGY USE INTENSITY (EUI)

RMI is an electricity-only building with no use of fossil fuels on site. The building is net-positive, generating more electricity than it uses. In terms of energy use intensity, this building is extremely efficient, with energy consumption well below the RIBA 2030 Climate Challenge benchmarks. The energy end-use bar shows that around 40% of consumption arises from heating and another 40% from appliance use with the remaining 20% is split between lighting and ventilation.

EUI	Measured by fuel	Measured by end use	
Renewables	-65	Renewables	-65
Heat	0	Heating and hot water	18
Electricity	47	Lighting	7
		Auxiliary power	5
		Appliance and plugloads	17

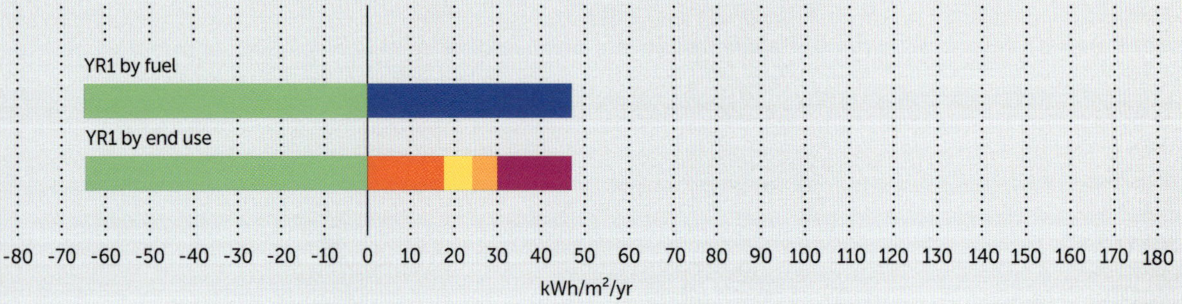

first year, who worked with the project team to validate performance by tracking data in real time and ensuring that control systems were operating as intended. The commissioning agent's role was critical for integrating the building's multiple interdependent systems, and to ensure that a platform for robust, ongoing real-time data monitoring was correctly set up.

Employees can manually control window opening, lighting, ceiling fans and the HyperChairs. A survey of occupants conducted at the end of year one found 100% satisfaction with daylighting and artificial lighting, and 88% satisfaction with thermal comfort in the building. Almost 80% said they spent more time in the office because they like the space. Another post-occupancy survey in 2018 found even higher levels of satisfaction among staff, with all aspects of the building ranking over 95%, except acoustics (90%).[5]

Energy monitoring is a mandatory part of IPD, and it continued for 18 months after completion, funded through savings on utility bills. Because the building utilised less energy than its predicted energy use intensity, the contractor and design team received a bonus.

With hindsight, crucial to the building's success was hiring an energy modeller and a commissioning agent. This highlights the importance of accurate predictive modelling and proper handover, even with a committed client and sophisticated design team.

ZGF architect Chris Chatto describes the role of feedback in achieving optimal performance at the RMI headquarters as 'absolutely critical, both because of the level of integration required to hit the high level of energy efficiency, and because occupants are so central to a passive building. Monitoring gives people the feedback they need to modify their actions. Furthermore, occupant surveys confirmed the success of passive thermal comfort strategies.'

KEY MESSAGES

- **Integrated design from concept stage through to feedback aligned performance targets and financial incentive across the project team.**
- **Integrated project delivery fostered a culture of collaboration.**
- **An adaptive comfort strategy permitted temperature swings within an acceptable range, reducing energy loads.**
- **Extensive metering and sensors facilitated gathering of post-occupancy data.**
- **Thorough commissioning ensured that systems operate as intended.**
- **Continuous post-occupancy monitoring enables rapid trouble-shooting.**
- **Staff includes both an on-site facilities manager and an engineer, who monitor seasonal adjustments, ensure ongoing performance in use, and troubleshoot and rectify problems.**

KEY FACTS

10.8% — INCREASE IN DESIGN TEAM FEES TO DELIVER ZERO CARBON

CONSTRUCTION COST UPLIFT TO DELIVER NET ZERO — $32/SQ FT

NUMBER OF SENSORS IN BUILDING — 120

NUMBER OF YEARS TO PAY BACK COST UPLIFT THROUGH ENERGY SAVINGS AND INCREASED PRODUCTIVITY — 3.9

NUMBER OF ELECTRIC VEHICLES POWERED BY THE BUILDING — 6

CASE STUDY 5
FREDERIKSBJERG SCHOOL

TYPE HYBRID MECHANICALLY AND NATURALLY VENTILATED SCHOOL

LOCATION AARHUS, DENMARK

COMPLETION 2016

CLIENT AARHUS CITY COUNCIL

GROSS FLOOR AREA 15,000 M^2

COST £31.8 MILLION

COST PER SQUARE METRE £2,100

NUMBER OF OCCUPANTS APPROXIMATELY 965 STUDENTS AND 105 TEACHERS AND STAFF

FORM OF CONTRACT DESIGN AND BUILD

ENVIRONMENTAL RATING DANISH BUILDING CLASS 2020 LEVEL (TOTAL ENERGY DEMAND FOR HEATING, VENTILATION, COOLING, DOMESTIC HOT WATER AND LIGHTING/M^2 HEATED FLOOR AREA LESS THAN 25 KWH/M^2/YR)

ARCHITECTS HENNING LARSEN ARCHITECTS AND GPP ARKITEKTER

ENGINEER NIRAS

LANDSCAPE ARCHITECT MØLLER & GRØNBORG

CONTRACTOR HOFFMANN

FIGURE 9.5.1
The school's south-facing facade features two sizes of windows (all fitted with automated external blinds with a manual override) located at different heights within the wall depending on internal layout.

FIGURE 9.5.2: OPPOSITE
An atrium at the heart of the building floods the interior with daylight and enables ventilation via the stack effect.

FIGURE 9.5.3
Cluster teaching area with amphitheatre and deep window reveals that double as benches.

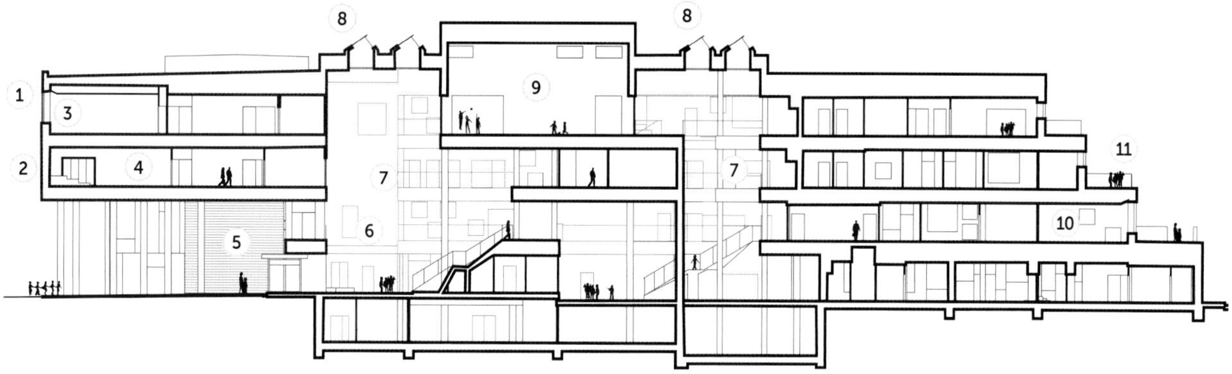

FIGURE 9.5.4
Section.

1 Automated external blinds with manual override
2 Stepped timber seating – learning stairs
3 Deep window cills with triple-glazed windows
4 Typical classroom
5 Double height forecourt
6 Climbing wall
7 Naturally ventilated indoor playground / central atrium space / circulation space
8 Automated roof lights
9 Sports hall
10 High and low level windows
11 Roof terraces / sport terrains

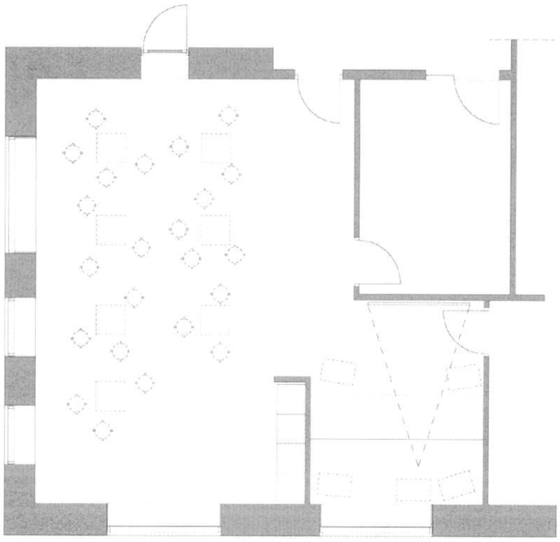

FIGURE. 9.5.5
Classrooms are divided into three zones: an open-plan area with tables, a small timber amphitheatre for group instruction, and a specialist teaching area subdivided by glazed partitions.

Located in central Aarhus, Denmark's second city, Henning Larsen's Frederiksbjerg School pioneered a new approach to Danish school design following a major national curriculum reform in 2014. The main thrust of the reform was to promote more active and collaborative learning, as well as more physical activity throughout the school day. The architects' competition-winning response to this complex brief is a highly articulated building with multiple entrances and rooftop play areas accessible to the community outside school hours. The programme includes a health centre, a daycare facility and a youth club; a mix of out-of-hours uses, which makes good zoning critical. Daycare is located on the ground floor, with classrooms for progressively older children as one moves up the building.

The new building, which houses approximately 965 students aged 3 to 16, is deliberately designed to encourage physical activity both inside the school and throughout the school campus. Roof terraces morph into sports terrains that morph into stepped seating for just hanging out. The entrance lobby opens into a four-storey atrium with a stepped platform, which doubles as assembly space and vertical circulation. Architects Henning Larsen describe the school as 'almost equal parts school and playground'.

Classrooms are organised in clusters, which include flexible open-plan teaching spaces, a specialist teaching area and a place for presentations with stepped timber seating, referred to as 'learning stairs', which serves as a miniature amphitheatre.

Design approach

Frederiksbjerg School's complex massing was developed to support active learning and increased physical activity, as well as to be accessible and welcoming to the surrounding community. The need for good daylighting also informed the design.

According to the architects, energy optimisation was not a primary design driver. Nonetheless, well-insulated external walls are readily apparent in the deep interior window cills and triple glazing, and an absence of thermal bridging reflects Denmark's stringent building regulations. According to the Danish Building Regulations 2010, the school meets Sustainability Building Class 2020, close to Passivhaus standard.

A central atrium enables natural ventilation via a stack effect. Both the atrium and the halls have a more dynamic thermal environment than a typical school, because the children are not just walking but are very active in these spaces – running, jumping, sliding, and skipping.

The position and size of the windows is a central element of the school's architectural expression. 'The way the elevation is composed corresponds very directly to the interior,' notes Henning Larsen partner Jakob Strømann-Andersen. Each classroom area has two sizes of windows located at three different heights within the wall: windows near the ceiling are intended to bounce daylight into the interiors, large windows are situated at eye level for both views and daylight, and low windows double as seating where children can snuggle up with a book. The thick walls were envisioned by the architects as *hygge* or cosy habitable nooks recessed in the walls.

All classroom windows except those that are north-facing have integrated external blinds that operate on sensors with a manual override. However, when blinds were in the closed position during our visit, the rooms were uncomfortably dark and lights were often turned on.

Feedback and actual performance

Frederiksbjerg School provides insights into how building performance is undertaken in Denmark, where a centralised team in Aarhus municipal government

FIGURE 9.5.6: OPPOSITE
Typical classroom fitted with uniform overhead lighting to deliver 300 lux at desk height. The variety of window sizes and locations can be seen in the wall beyond.

FIGURE. 9.5.7
Students gravitated to small group teaching around the tables fitted with the new pendants.

collects energy data from all schools and is able to make comparisons that identify schools with high operational energy use.

Researchers at Aarhus University have undertaken indoor environmental quality (IEQ) post-occupancy studies to monitor the school's indoor thermal climate, air quality and lighting. The building performs well in terms of indoor climate, and interestingly, CO_2 measurements were used not only to check acceptable levels but to detect occupancy patterns within different rooms in order to optimise the use of rooms during the school day and for after-school activities. For example, by monitoring CO_2 levels, the study found that rooms set aside for cooking instruction were underused, while the demand for music practice rooms outstripped supply; music practice was subsequently relocated to the underused rooms.

According to the technical staff, the school has significantly exceeded design-stage predictions for both occupancy and hours of operation. While this is a reflection of the success of the building, it makes it difficult to compare predicted performance and actual energy use, highlighting that in the planning of the feedback process it might be necessary to develop multiple scenarios.

Mirroring experience in the UK, initial energy monitoring revealed that energy consumption was higher than predicted. Recommissioning was undertaken, significantly reducing operational loads.

Qualitative feedback received while visiting the school suggests that Frederiksbjerg School is a highly stimulating environment, perhaps overly so. One comment was that the environment did not help students maintain focus. In one classroom, a teacher had applied a translucent film to the glazed walls of the group work area to create privacy.

Post-occupancy lighting study

Lighting feedback studies[6] found that the quality of daylighting varied widely throughout the school and was dependent on the placement of windows in the facade, particularly the height of the window in the exterior wall. Windows placed in the lower part of the exterior wall did not provide adequate daylighting at desk height. Because of the window placement and depth of the classrooms, electric lighting was often needed at all hours of the day.

Occupancy feedback also found that the combination of an open flexible plan with uniform overhead lighting meant that the classrooms lacked a focal point and enclosure. The architects worked with academic partners to understand and address this concern.[7] What they realised was, that while considerable attention was paid to window placement at design stage, a one-size-fits-all approach to electric lighting was adopted. LED fixtures with daylight-responsive dimming were evenly positioned across all classroom ceilings to deliver a uniform

DIFFERENTIATED LIGHTING SIMULTANEOUSLY IMPROVED CONCENTRATION AND REDUCED ENERGY BILLS BY UP TO 68%.

300 lux at desk height, as required by Danish building regulations.

To test whether introducing a greater variety of lighting would enhance learning and also save energy, pendant light fixtures were introduced to two classrooms to define work areas by creating pools of light. The premise was to explore whether variation in lighting would enhance students' comfort and performance. Using personal observation, videos, interviews, sound recordings, and actual measurements of lux levels and energy usage, researchers conducted behavioural studies, light mapping, movement, acoustic and academic exercises to assess the impact of the pendant lighting on students and teachers. Activity in the two classrooms

FIGURE. 9.5.8 FREDRIKSBJERG SCHOOL ENERGY USE INTENSITY (EUI)

Fredriksbjerg School energy use intensity data shows that even buildings with generous spaces and complex form can have a low energy demand, with the total energy use intensity approaching the RIBA 2030 Climate Challenge benchmarks for non-domestic buildings. Heating consumption was somewhat higher than expected, and this data set is prior to the building heating systems undergoing recommissioning. Even then, heat consumption is within the typical UK range, and electricity use is below the CIBSE good practice benchmark.

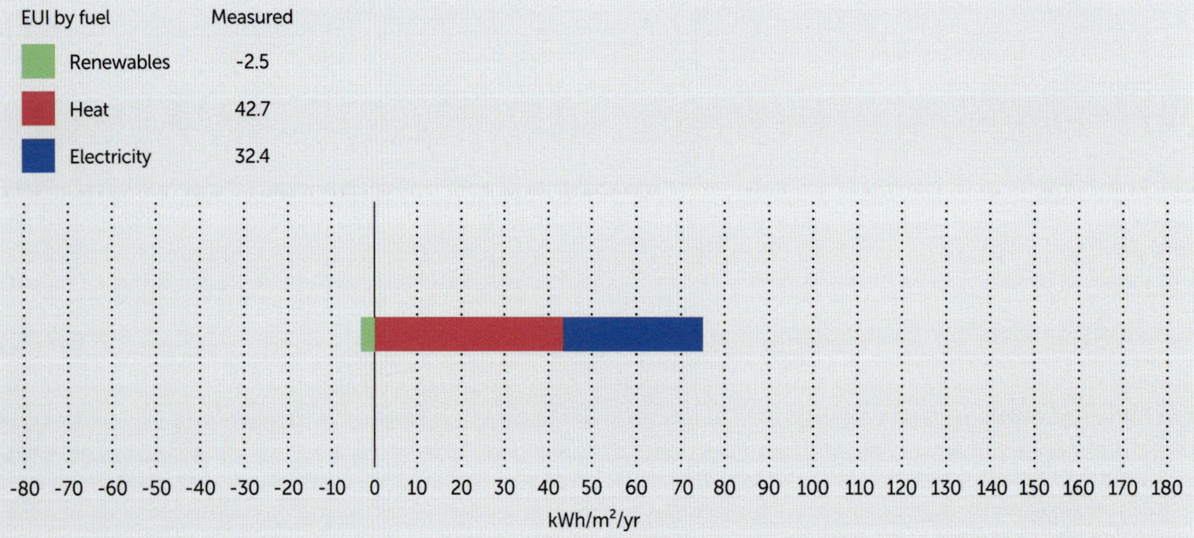

with pendant lighting was compared to two identical classrooms without the pendants.

The pendant lighting solution was found to nurture quietness and help more active children settle, with children actively gathering around the pendant lights. Measured light and decibel levels found significantly reduced noise levels after the introduction of the pendant lights. In 70% of cases, noise was reduced by 1.7 dB.

The researchers concluded that 'pupils were attracted to lighting they intuitively associate with the safe and comfortable atmosphere of their home decor', and that the pendant lighting encouraged peer-to-peer learning rather than them always relying on the teacher. Differentiated lighting, particularly illuminated zones at eye level, simultaneously improved concentration and reduced energy bills up to 68%.

Regulatory changes

To date, Denmark's 2014 curriculum reform has not been accompanied by revised ventilation and lighting standards for schools. An approach which incorporates greater adaptive comfort criteria would suit an open-plan active learning environment such as Frederiksbjerg School.

In 2017, following the completion of Frederiksbjerg School, Danish building regulations were amended to require a functional test of a building's technical systems, (including HVAC and lighting), before a new or refurbished building is commissioned. While the test is mandatory and the results must be submitted to the municipality, it is recommended, but not required, that it be undertaken by a third party.[8] Mandated testing of building systems, if properly enforced, could contribute to closing the performance gap by addressing the many issues with systems and controls described in chapters 7 and 8.

KEY MESSAGES

- **Centralised energy monitoring by the City of Aarhus facilitated data collection and troubleshooting and enabled the school to recommission its systems to reduce energy use.**
- **CO_2 monitoring was used innovatively to detect occupancy levels in different rooms and optimise activities and room occupancy across the school.**
- **Installation of pendant lighting in two test classrooms improved the teaching environment by creating calmer areas for concentration, but the school has yet to implement this approach more widely.**
- **Working with academic partners enabled detailed studies and lessons learned for future projects.**

KEY FACTS

REDUCTION IN ENERGY BILLS DUE TO TRIAL INTRODUCTION OF PENDANT LIGHTS **68%**

GLAZING RATIO **33%** OF FACADE AREA

650 mm THICKNESS OF EXTERNAL WALL INCLUDING 350 mm OF INSULATION

REDUCTION IN CLASSROOM DECIBEL LEVELS DUE TO INTRODUCTION OF PENDANT LIGHTS **1.7 dB** IN 70% OF CASES

400,000 EXISTING BRICKS REUSED, SAVING 180 TONNES CO_2

CASE STUDY 6
KEYNSHAM CIVIC CENTRE, LIBRARY & INFORMATION SERVICE

TYPE NATURALLY VENTILATED OFFICE (PUBLIC SECTOR)

LOCATION KEYNSHAM, SOMERSET

COMPLETION 2014

CLIENT BATH AND NORTH EAST SOMERSET COUNCIL

GROSS FLOOR AREA 6,325 M^2

COST £23,060,000 (EXCLUDING EXTERNALS AND HIGHWAY WORKS)

COST PER SQUARE METRE £3,645

NUMBER OF OCCUPANTS APPROXIMATELY 455

SERVICES HEAT PUMPS SUPPLEMENTED BY GAS-FIRED CONDENSING BOILERS SUPPLY UNDERFLOOR HEATING AND PERIMETER TRENCH HEATERS, AND COOL SERVER AND MEETING ROOMS; NATURALLY VENTILATED

AIRTIGHTNESS 4.7 $M^3/HR/M^2$ @ 50 PA

FORM OF CONTRACT NEC DESIGN AND BUILD

ENVIRONMENTAL CERTIFICATION NONE OPTED FOR

ENVIRONMENTAL RATING DEC B/36 (2017)

DEC B/36 (2017)

ARCHITECT AHR

SERVICES ENGINEER MAX FORDHAM

STRUCTURAL ENGINEER HYDROCK

LANDSCAPE ARCHITECT NOVELL TULLETT/AHR

CONTRACTOR WILLMOTT DIXON

A focus on sustainability performance, not just aspirations, was central to the Keynsham Civic Centre project from the outset. Architects AHR had recently undertaken seven Innovate UK-funded BPE studies that explored in detail the many reasons why buildings fail to perform as designed. To avoid the common pitfalls that cause the performance gap, the client – together with AHR and engineers Max Fordham – structured the design and delivery process for the Keynsham project differently. Energy was considered at every step, and the project takes an innovative approach to delivering exemplary building performance.

Design approach

The team pioneered several measures, including an operational energy-use budget and an energy risk register (see Building Performance Register, chapter 10), both incorporated into a performance contract that

FIGURE 9.6.1
Site plan.

FIGURE 9.6.2 : OPPOSITE
Entrance overlooked by open-plan offices.

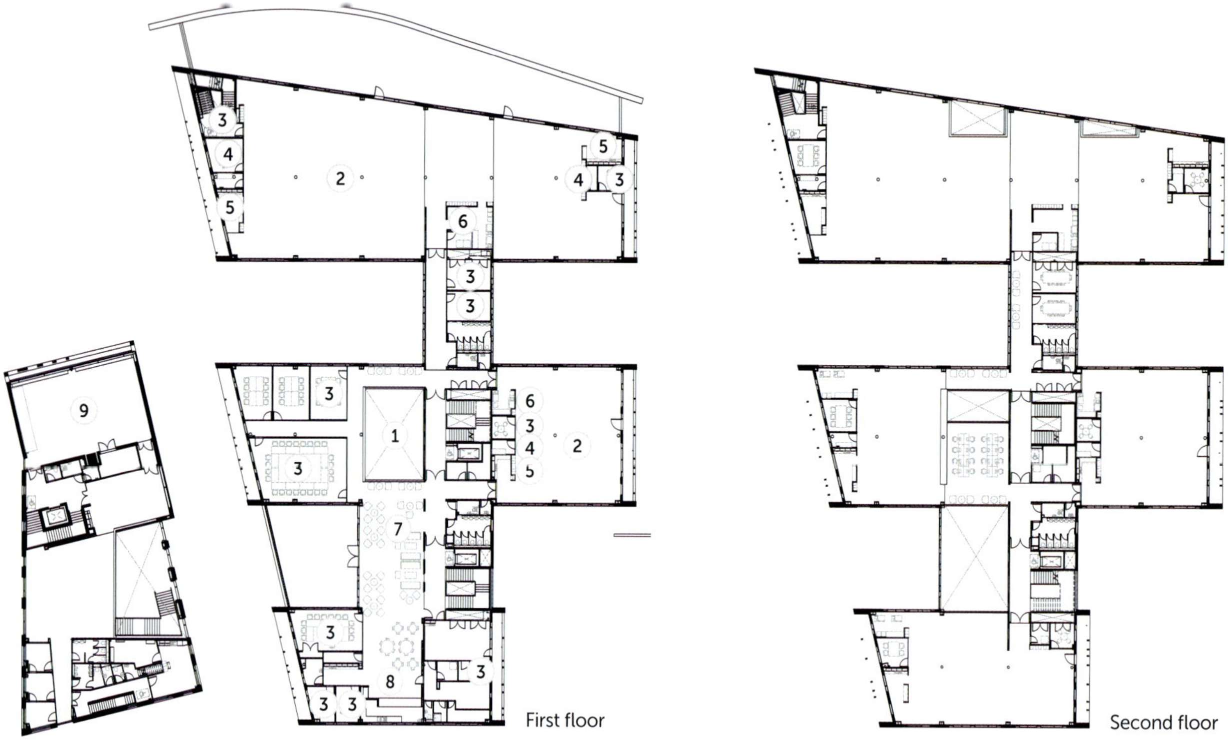

FIGURE 9.6.3
First and second floor plans.

1 Light well over entrance
2 Open plan office space
3 Meeting room
4 Confidential room
5 Tea point
6 Copy area
7 Breakout space
8 Cafeteria
9 'Multi purpose meeting & performance space

targeted a Display Energy Certificate (DEC) A rating two years after project completion.

Initiated by the architects, an 'energy efficiency measures register' listed every aspect of the building that could impact future energy use, from floor-to-floor heights, to thermal mass, to window systems. As the project progressed, the Soft Landings (SL) framework was chosen as the process to meet the DEC A target, with Max Fordham acting as the client's SL champion from design through to building aftercare. Max Fordham adopted the energy register and operational energy budgets and incorporated them into the contractor's preliminaries, thereby making performance contractually binding and giving the SL process real teeth. This ensured that if any changes were made that impacted performance or energy consumption, the energy model had to be re-calculated by the contractor to demonstrate that overall performance would not be jeopardised.

According to the engineers, the requirement to complete the design within the energy budget resulted in fewer changes and less dilution of the original design than is common in most design and build contracts.

The contractor was responsible for tracking energy performance against the baseline energy budget during the first two years of operation, reporting and benchmarking consumption against the calculated baseline on a monthly basis to the client. This approach shifted the conventional incentives of project delivery and resulted in a more collaborative process with better outcomes.

Detailed 'pre-occupancy' studies of the existing council offices had documented staff working habits and needs, as well as energy use patterns. The facilities management team were also interviewed for better understanding of operational needs, and to ensure that management and maintenance concerns

FIGURE 9.6.4
Approach from northeast.

were taken on board early. This information was drawn on to establish an 'energy budget' for the new premises based on actual occupancy patterns and energy needs, rather than relying on a notional compliance model (see chapter 1). The energy budget was agreed upfront and later incorporated into the contractor's design and build contract.

The brief outlined requirements for a new civic centre which would be both flexible in use and resilient to climate change. A major driver was to reduce the high operating costs of the previous premises. The complex brief included a library, a one-stop shop and 1850 m^2 of retail, in addition to 6300 m^2 of council offices, to be achieved on a challenging site.

The architects responded by creating a public square fronting the high street. The civic centre itself is bisected to create a landscaped pedestrian link to the park below.

Passive design informed the building's massing and east–west orientation. Three 15-m deep wings – deliberately narrow for cross-ventilation – are topped by a sawtooth roof that provides deep daylight penetration via a north-facing clerestory, and stack-effect ventilation. Roofs are oriented for optimal solar gain, and house 1150 m^2 of photovoltaics for on-site electrical energy production.

Keynsham Civic Centre's structural system is an exposed cross-laminated timber (CLT) frame. This approach offered the double benefit of speeding up construction and reducing the building's embodied carbon compared to conventional concrete or steel construction. Innovative precast concrete panels with cooling pipework cast into the concrete comprised 50% of the floors and ceilings, increasing the thermal mass.

A central atrium allows views into different departments to encourage better communication

FIGURE. 9.6.5
Second floor offices with clerestory windows.

and collaboration, while ceiling heights of just over 3 m make for inspiring workspaces and mean that in summer warm air can stratify, while the zone people occupy remains cooler.

A bespoke window system consists of acoustically attenuated louvred panels that filter out noise from surrounding streets and can be manually operated for ventilation, while automated high-level windows allow for night purging of warm air. Light shelves bounce natural light into the interiors and are supplemented by manually operated blinds to control glare.

The project was honed at the design stage to maximise passive design and minimise technical systems, such as mechanical ventilation. The building's section, with an internal atrium and vent voids, meant that dedicated ventilation risers could be eliminated. Task lighting was provided at desks, with automated light detection in circulation zones only.

Feedback and actual performance

The monthly reporting of energy data exposed M&E issues such as installing, calibrating and linking systems to the BMS, and profiling heating systems and motorised vents.

The performance contract conditions enabled problems to be identified and mostly resolved as part of the aftercare process. Any issues with meters were fixed early on so that energy data could be regularly reported as per the contract conditions, and commissioning of systems was subject to a level of scrutiny and adjustment that few projects can afford post-completion. The requirement for the contractor to collect and benchmark data monthly was critical, in order to expose issues with metering, heating, ventilation and lighting, and enabled the contractor to make remarkable improvements while the project team was still engaged.

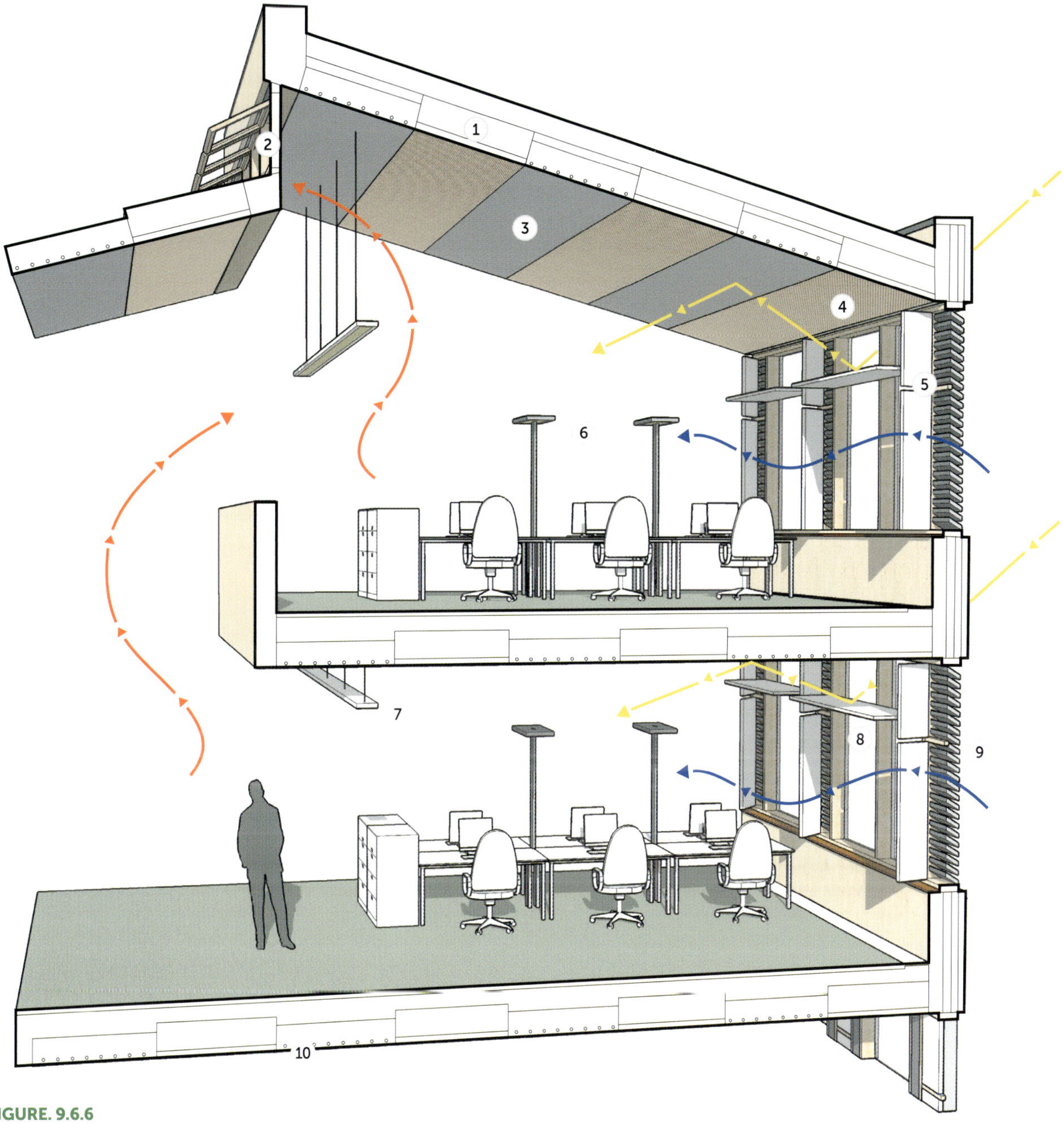

FIGURE. 9.6.6
Environmental section.

1 Hybrid cross-laminated timber (CLT) and steel frame structure

2 Clerestory windows exhaust warm air at roof level

3 High thermal mass concrete slabs with embedded cooling pipework make up to 50% of ceiling and floors

4 Perforated spruce panels control reverberation

5 BMS-controlled automated upper ventilation panel and manual lower panel

6 Task lighting with manual override at desks

7 High level automatic lighting to circulation routes only

8 Light shelves to south windows bounce light deeper into offices; manual blinds control glare at fixed glazing

9 Acoustic louvres absorb low-frequency traffic noise and allow secure night-time heat purge

10 Exposed soffits with raised access floor. Kill switch enables shut down of PCs and lights at night to save power

In a departure from conventional practice, an independent commissioning manager was appointed directly by the contractor, rather than by the M&E contractor. The Soft Landings process was strengthened by the contractual requirement for metrics, and the holistic scrutiny of architectural, services and operational evolution of the project through the performance register ensured that passive design features escaped value-engineering.

The aftercare process identified heat pump commissioning issues, and also a problem with the motorised vents for the clerestory windows, which had been programmed to open larger banks of windows than intended, explaining why staff had complained about cold draughts – this issue crops up repeatedly in feedback studies (see chapter 8).

A further research study into the energy performance and indoor environmental quality led by University College London found that the CO_2 levels and summertime temperatures on the top floor were unexpectedly high, due to faulty sensors. Because the performance contract for the project only required the monitoring of energy consumption, but not indoor air quality or occupant comfort, this had not been picked up by the aftercare process. An occupant survey with an 85% response rate found that employees were generally satisfied with indoor conditions, with the exception of minor acoustic issues common in open-plan offices.

Another issue which recurs in feedback studies and was found at Keynsham is that office workers tend to inadvertently leave blinds partially closed due to glare. This compromises the effectiveness of the light shelves and often triggers unnecessary use of

FIGURE. 9.1.8 KEYNSHAM CIVIC CENTRE ENERGY USE INTENSITY (EUI)

Due to the performance contract linking energy use reporting to contractual incentives, energy end-use data is available for calculated as well as operational records for every year since the building's completion. Comparison of data from the first two years pinpoints exactly where consumption was higher than predicted in year one, and the substantial impact of aftercare and remedial action. Aside from equipment and appliance loads (IT, catering and lifts), lighting, fans and pumps were all brought back to the low predicted levels. Heating consumption remains higher than expected because the heating system ran on the back-up boilers rather than the low-carbon heat pumps. When taking into account the energy generated by on-site renewables, this building is very close to meeting DEC A and RIBA 2030 targets.

EUI by end use	Calculated	Measured Y1	Measured Y2
Renewables	-31	-35	-30
Heating and hot water	19	37	29
Lighting	5	28	11
Auxiliary power	2	10	10
Appliance and plugloads	11	12	16

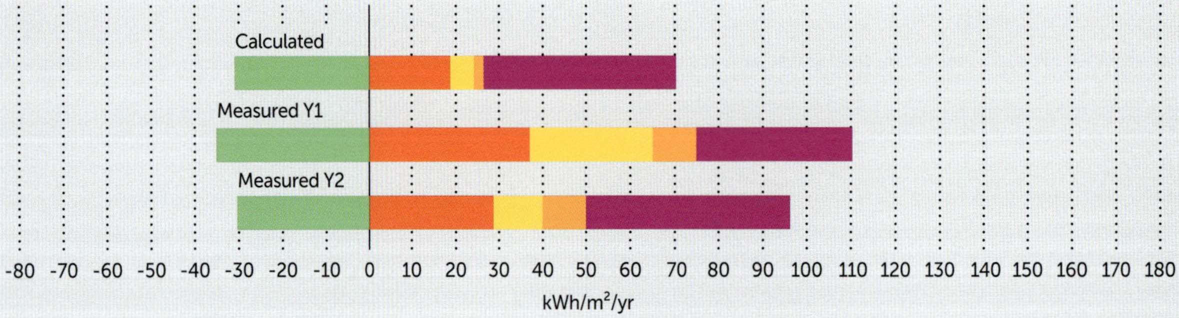

FIGURE. 9.6.7
View through the central spine of the building.

artificial lighting. The manual windows also caused some initial problems in practice. For example, someone from the FM team must ensure that no panels are left open overnight.

The contractor, together with the council's FM team, compiled a detailed, 35-page self-explanatory user manual for council staff, primarily composed of photos with short captions. Design and FM teams worked with building users on a regular basis – and seasonally – during the first years of occupancy to solve problems as they arose, and new employees are thoroughly briefed on the building's operation. To control unpredicted energy loads by occupants, clear restrictions on the use of electronic equipment such as desk fans, heaters or kitchen appliances were established.

Keynsham Civic Centre illustrates how the public sector can pave the way to net zero performance through exceptional leadership, by pioneering innovative design and contractual processes and widely sharing data and lessons learned.

KEY MESSAGES

- **Precast concrete panels in floors and ceilings add thermal mass to the lightweight building.**
- **Contractually binding energy budget with DEC A target required that performance outcomes be tracked throughout design, procurement and construction.**
- **Contractor required to benchmark data monthly against energy budget from day one.**
- **Mandatory adoption of Soft Landings and comprehensive contractual requirements outlining specific roles and responsibilities for aftercare ensured that BPE was embedded in procurement.**
- **Proactive engagement with the FM team and building occupants helped ensure that users' needs were addressed and that occupants understood how to operate the building.**

KEY FACTS

96% REDUCTION IN ENERGY COSTS COMPARED TO PREVIOUS SITES

15% ENERGY SAVINGS IN 2ND YEAR OF OPERATION VS 1ST YEAR

50% LOWER FOSSIL ENERGY USE COMPARED TO CIBSE BEST PRACTICE[9]

TOTAL ANNUAL CO_2 EMISSIONS/m^2 GIFA
$15.2\ kgCO_2/m^2/yr$

90% TOWARDS THE RIBA 2030 ENERGY CONSUMPTION TARGET IN 2017

CASE STUDY 7
ENTERPRISE CENTRE, UNIVERSITY OF EAST ANGLIA

TYPE NATURALLY VENTILATED OFFICE AND CLASSROOM BUILDING (PASSIVHAUS)

LOCATION NORWICH, NORFOLK

COMPLETION 2015

CLIENT UNIVERSITY OF EAST ANGLIA AND THE ADAPT LOW CARBON GROUP

GROSS FLOOR AREA 3400 M^2

COST £11.6 MILLION

COST PER SQUARE METRE £3,412

NUMBER OF OCCUPANTS 100–300

SERVICES MIXED-MODE, WITH DEMAND-CONTROLLED MECHANICAL VENTILATION WITH VAV DAMPERS TO EACH OCCUPIED SPACE AND HIGH-EFFICIENCY AHUS; HEATING SUPPLIED VIA HEAT EXCHANGER LINKED TO UNIVERSITY DISTRICT HEATING; 50 KW PV ARRAY MEETS APPROXIMATELY ⅓ OF ELECTRICAL LOADS

AIRTIGHTNESS 0.2 M^3/HR/M^2 @ 50 PA

FORM OF CONTRACT SINGLE-POINT DELIVERY (SPD) FOR ALL WORK STAGES (BASED ON NEC3)

ENVIRONMENTAL CERTIFICATION PASSIVHAUS, BREEAM OUTSTANDING

ENVIRONMENTAL RATING DEC A

ARCHITECT ARCHITYPE

SERVICES ENGINEER BDP

STRUCTURAL ENGINEER BDP

LANDSCAPE ARCHITECT CHURCHMAN LANDSCAPE ARCHITECTS

CONTRACTOR MORGAN SINDALL

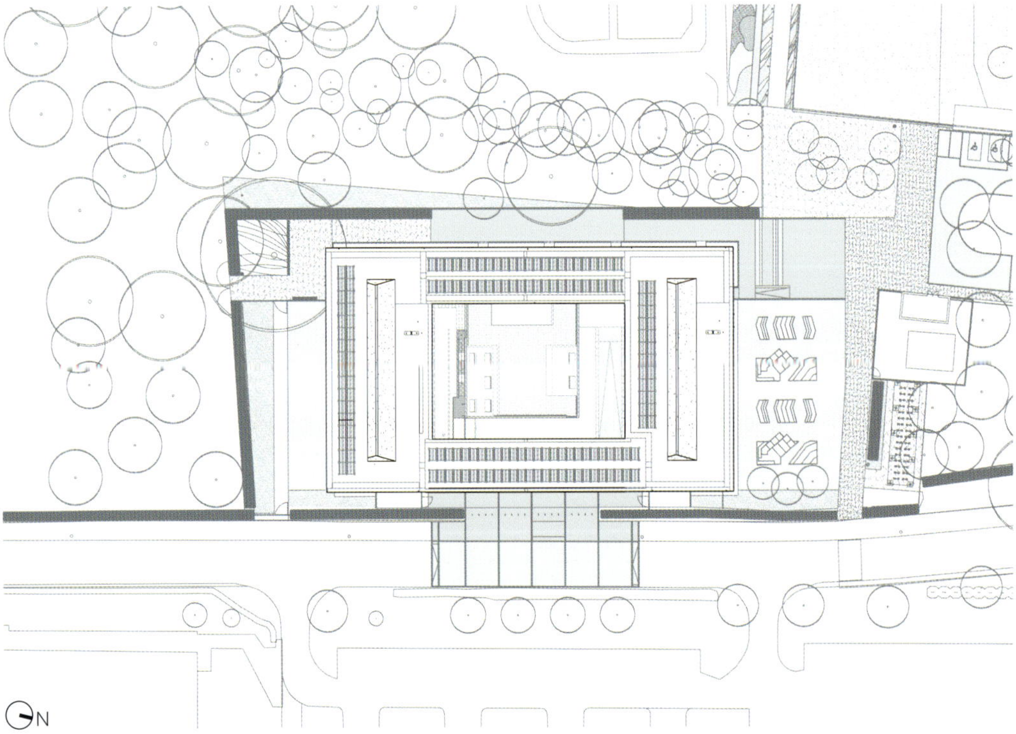

FIGURE 9.7.1
Site plan.

FIGURE 9.7.2: OPPOSITE
South elevation, with brise-soleils and louvred ventilation panels that provide fresh air without compromising security.

FIGURE 9.7.3
Main entrance, with classroom wing to the south (left) and office incubator space to the north (right).

The Enterprise Centre at the University of East Anglia (UEA) in Norwich stands out because of its approach to reducing both operational and embodied carbon, and its rigorous use of low-carbon, locally sourced materials. This is a one-off building commissioned by an ambitious client, UEA and the Adapt Low Carbon Group, with a proactive estates department that monitors data from every building on the campus. Much can be learned from this project in terms of feedback that applies across other building types.

The project brief called for seminar rooms, a variety of office spaces for start-up businesses, an auditorium and an events space, and targeted Passivhaus and BREEAM Outstanding certifications. Further requirements included a 100-year life span, an ambitious threshold for embodied carbon, and 70% nature-based materials.

Design approach

The Enterprise Centre is a courtyard building, with the most densely occupied office spaces to the north and intermittently occupied classrooms facing south. The exhibition and events space linking the office and classroom wings faces west, and is designed to withstand dynamic changes in temperature and light levels. These three wings wrap around the auditorium, partially below ground, which occupies the heart of the site.

Office and classroom wing corridors feature a clerestory for stack ventilation – the only windows in the building linked by actuators to the BMS system, because they are out of reach. High ceilings enhance spatial quality and allow stratified hot air to remain above the occupied zone and be extracted through the clerestory windows without causing occupant discomfort. Radiant heat from lighting is also emitted into this upper zone.

All remaining windows are manually operated,

THE ENTERPRISE CENTRE AT THE UNIVERSITY OF EAST ANGLIA IN NORWICH STANDS OUT BECAUSE OF ITS APPROACH TO REDUCING BOTH OPERATIONAL AND EMBODIED CARBON, AND ITS RIGOROUS USE OF LOW-CARBON, LOCALLY SOURCED MATERIALS.

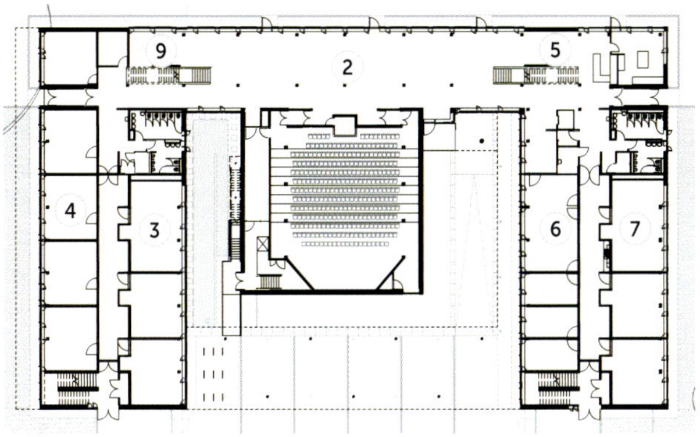

Ground floor

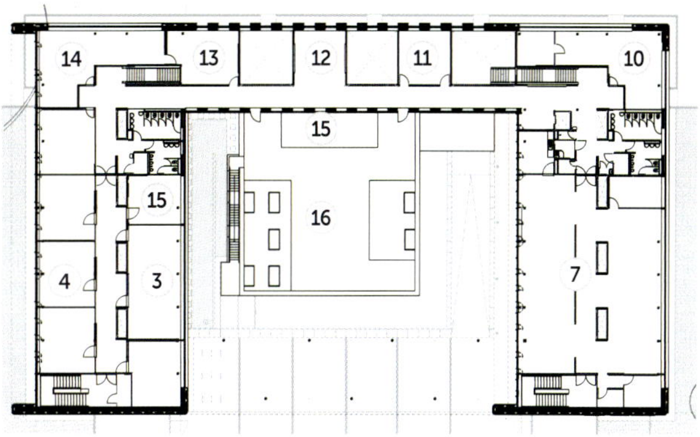

First floor

N

FIGURE 9.7.4

Ground and first floor plans.

1 Lecture theatre
2 Exhibition
3 Lecture / seminar
4 Adaptable seminar
5 Cafe
6 Student enterprise
7 Hatchery
8 Gallery
9 Social work
10 Office
11 Board room
12 Bar
13 Lab
14 Breakout space
15 Plant
16 Roof terrace

and every habitable room has an openable window. Windows on the east, west and north facades are sized to the minimum required to achieve required lux levels, while deep overhangs on the south facade allow winter sun to enter the building while controlling glare throughout the year. To prevent overheating, the south facades feature translucent glazed panels which open on to fixed louvres to allow ventilation without compromising security. Guidance on when windows should be opened and closed was carefully considered. A small panel of window-opening guidance, fitted next to the temperature, humidity and CO_2 sensors in each room, went through several iterations until it was distilled into simple wording, with green and blue lights which illuminate when windows should be kept shut or may be opened. A red light was deliberately avoided, because it might suggest a fault.

An exposed concrete ground floor, an additional screed at first floor, and higher-density linings to all spaces add thermal mass to the lightweight building to reduce temperature swings. A well-insulated airtight envelope further contributes to stable interior temperatures, reducing demand for heating and cooling. The exterior walls are comprised of a Larsen truss, in which non-loadbearing oriented strand board (OSB) fins are attached to the loadbearing exterior wall to create a 200-mm wall cavity for insulation without thermal bridging.

From the outset, the design team set out to reduce all energy demand, both regulated and unregulated (see chapter 3). The offices have no ceiling lights, and individual task lighting is provided at each desk. Due to Architype's experience with small power loads exceeding anticipated levels on previous projects, significant time was invested early to work with suppliers of AV and IT equipment to reduce these loads to a minimum.

In occupied rooms, heating is provided by small radiators controlled by the BMS. Only the server

FIGURE 9.7.5
In the corridors of both the classroom and office wings, north-facing clerestory windows on actuators draw air out of the building through the stack effect.

room and the auditorium have mechanical cooling. Routes for supply and extract air were designed into the building during RIBA stage 2,[10] which means that a duct loop supplies each wing of the building rather than individual rooms, minimising duct runs and bulkheads. A small PV array meets approximately 32% of electrical loads.

The university opted for an NEC3 partnering contract, enabling a highly integrated design approach because contractor and design team are appointed under a single contract and hence work together at an agreed price. This enabled substantial involvement of subcontractors in design development. Uniquely, the architects were involved in interviewing the subcontractors for thatching, glulam structure and M&E.

The architects also reviewed the operating manual for the BMS system, ensuring that set points and timings were appropriate – a task Architype routinely undertakes on most jobs.

Feedback and actual performance

Architype undertook three years of structured post-occupancy evaluation, as agreed from the outset as part of the Soft Landings process. This included monitoring energy and internal comfort data as well as two BUS surveys and a thermal imaging report, all undertaken by BSRIA. The thermal imaging report was used during snagging. A reserve (between 1% and 2% of the contract sum) set aside at contract stage to resolve post-completion problems was critical to the success of this process.

Because monitoring was planned for from the beginning, appropriate metering was in place. An initial year of adjusting and fine-tuning building operation was followed by a second year of monitoring and the initial BUS survey. Further remedial actions were followed up by a second BUS survey in year three.

Certified as Passivhaus following completion, the Enterprise Centre has maintained its DEC A rating for

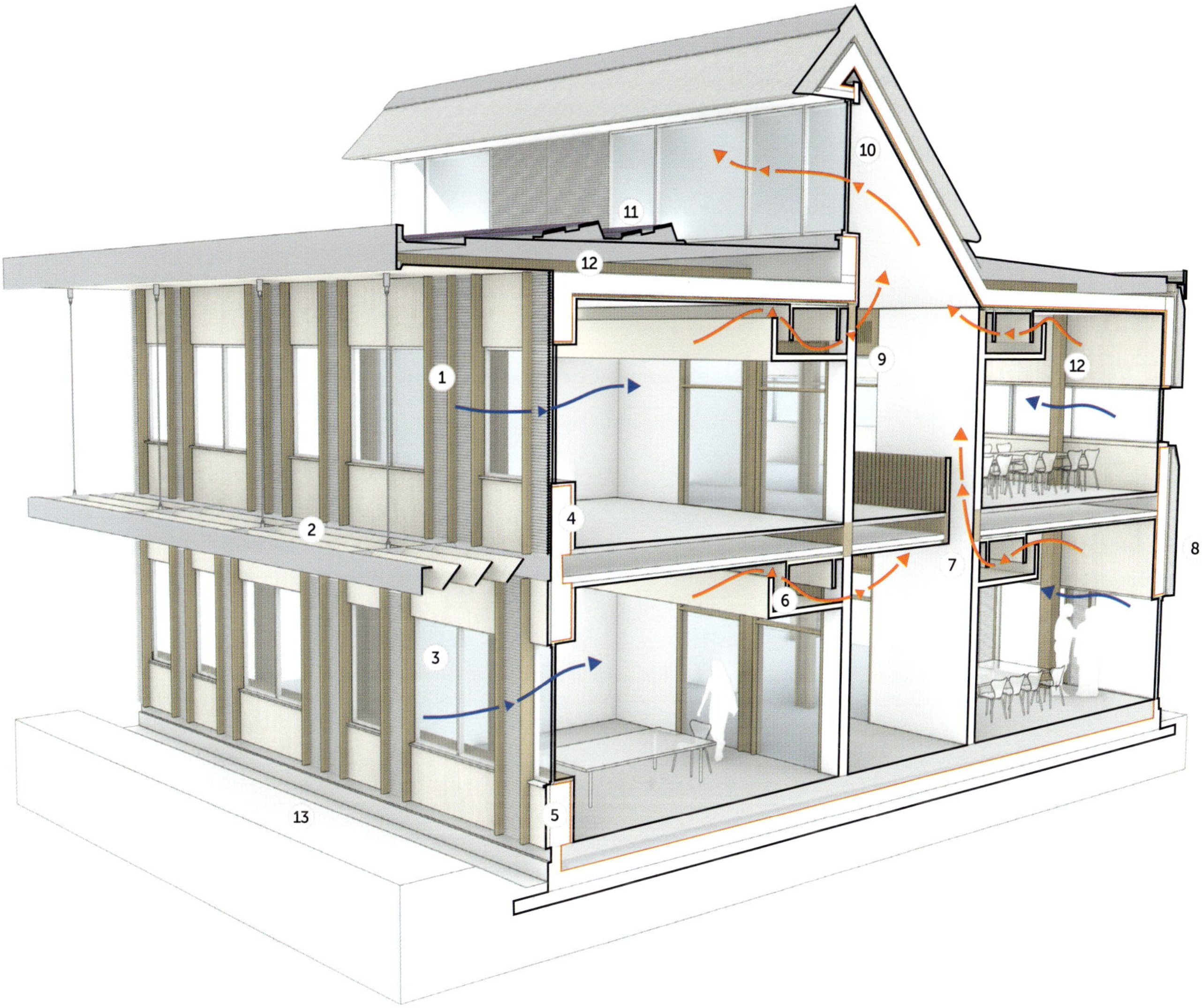

FIGURE 9.7.6

Environmental section.

1 Facade grills with operable indoor panels for ventilation and secure night purge in summer

2 External shading hung from roof structure

3 Triple-glazed windows

4 Continuous airtightness layer detailed around the building envelope

5 Insulation wrapped around the building in a duvet layer

6 Accessible services zone and ventilation plenum

7 Corridor vent voids exhaust warm air to roof vents

8 Thatch cladding to walls and clerestory roofing

9 Grilles above doors exhaust used air into corridor

10 Clerestory windows on actuators

11 Low angle rooftop photovoltaic panels

12 Glulam frame with locally sourced timber for partitions and roof joists

13 Exposed concrete ground slab on insulated shuttering system for thermal mass

all four years despite significantly increased usage, particularly on the conference and catering fronts. Heating loads remain consistently below predicted levels for all four years; however, electricity loads have increased due to the building's heavy use.

During the Soft Landings period, the BMS automatically triggered an email to both architect and services engineer if temperatures exceeded 27°C or dropped below 18°C, or if CO_2 levels surpassed 1500 ppm or dropped below 400 ppm. This enabled prompt troubleshooting, including resetting the opening of the clerestory windows and rectifying a CO_2 sensor packed with sawdust.

Soft Landings

Gareth Selby, the project architect, describes the feedback process as an 'eye-opener'. The crux of post-occupancy evaluations is avoiding blame and rectifying problems. Earmarking a small percentage of the contract fee early to rectify post-completion issues helps circumvent the reasons why project teams are so often reluctant to undertake performance evaluation and share the findings.

Quarterly user meetings took place for three years after completion. The format and frequency of meetings and the attendees had all been agreed upfront. Selby notes the importance of scheduling these meetings early to ensure continued participation

FIGURE. 9.7.8 ENTERPRISE CENTRE ENERGY USE INTENSITY (EUI)

The Enterprise Centre is a rare example of a building where measured consumption in year one was lower than calculated. The year-on-year energy bars show that as the building settled in, consumption rose only marginally to within a few kWh/m²/yr of predictions, despite heavier-than-anticipated use.

Energy end-use data for year two shows the low heat consumption achievable with Passivhaus fabric performance and MVHR, even with fan power taken into account. Lighting, fan power and appliance loads are also low, nearly all in the single digits.

EUI by fuel	Calculated	Measured Y1
Renewables	-11	-4
Heat	17	15
Electricity	27	24

EUI by end use	Measured Y2
Heating and hot water	17
Lighting	6
Auxiliary power	10.8
Appliance and plugloads	11
Cooling	0.2

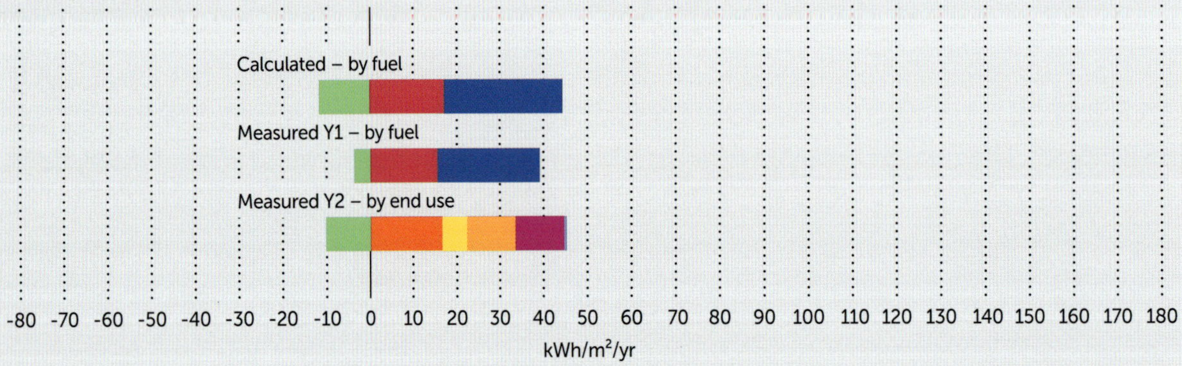

of project team members familiar with the building.

A problem which surfaced immediately was the faulty operation of light fittings in the offices that switched on and off randomly, causing many complaints from staff. Funds from the reserve budget were used to arrange a site meeting between client, architect, design engineer, supplier and BMS engineer. Due to all parties being present on site and a willingness to openly discuss the issue, the problem was quickly identified as a conflict between lighting controls and the building BMS. It was relatively easy to take remedial action, because the earmarked reserve was used to fix the problem, quickly resolving an issue without escalation.

Occupant surveys

In the two BUS surveys, conducted a year apart, the Enterprise Centre ranked at 95% in terms of overall performance among buildings in the BUS database. In terms of internal comfort, it ranked at 75%, well above the median benchmark. The two areas of concern highlighted by occupants were insufficient artificial lighting and cool indoor temperatures, both in winter. Because sensors were located in every occupied room, it was also easy to check overheating, which occurred in two rooms where occupancy was higher than anticipated.

Due to evolving client needs, at the time of writing Architype is exploring a range of alterations to the building that may also enhance performance. 'Technology has evolved in five years, and we're looking at replacing lighting and hoping to make a good building even better,' explains associate Christian Dimbleby.

In terms of benchmarking, the Enterprise Centre is a 70 $kWh/m^2/yr$ building, marginally exceeding the RIBA 2030 Climate Challenge's target of 55 $kWh/m^2/yr$. The fact that this ambitious low-carbon building should perform better still highlights the extent of change required to deliver zero carbon buildings and the important role of feedback on this journey.

KEY MESSAGES

- **Even with an ambitious client and project team that championed sustainability at all stages, proper commissioning and follow-up were essential to troubleshoot problems and ensure good performance.**
- **A contractual arrangement enabled architects to interview and engage with subcontractors early, as well as setting aside a small percentage of the contract fee to rectify post-completion issues.**
- **A rigorous approach to fabric performance and solar shading permeate the design.**
- **Regulated and unregulated loads were scrutinised from the design stage so that all energy end uses were accounted for in the energy budget.**
- **Soft Landings was written into the contract from the outset so that post-occupancy monitoring was planned for, appropriate metering was in place, and a three-year aftercare process with pre-scheduled quarterly meetings was agreed.**

KEY FACTS

33% ELECTRIC LOADS MET BY PVS

100 YEAR LIFE SPAN

70% BIO-BASED MATERIALS

CONTRACT SUM FOR FEEDBACK PROCESSES <1.5%

3 YEARS OF POE VIA THE SOFT LANDINGS FRAMEWORK

10

CONTRACT FOR PERFORMANCE

FIGURE 10.1: OPPOSITE
Rocky Mountain Institute, Basalt, Colorado.

Introduction

The previous chapters detailed the transformative role that feedback can play in delivering zero carbon buildings. To be effective, a clear process is needed to establish feedback targets at the outset of a project and track them through to building operation. Every project team needs to determine which feedback metrics are most relevant for a particular building and agree early in a project cycle how the information will be gathered, monitored and reported. Responsibility for each deliverable has to be allocated and risks managed.

Procurement is the critical leverage point for incorporating effective feedback processes into projects. This chapter explores in detail contractual mechanisms for including and validating performance targets. It sets out the use of a Building Performance Register (BPR) as a tool that makes it possible to track and share building performance parameters across the supply chain so that they are not diluted over the life of a project. A step-by-step approach for different stages of a design project is mapped against the RIBA Plan of Work.

Building Performance Register

A Building Performance Register (BPR) is a record used to track all the elements of a building that impact energy performance at key milestones of a project, and how these milestones are going to be achieved, unaffected by value-engineering. This allows information that is critical for successful performance to be easily shared across a project team. While a BPR can be managed by any member of the design team, for most projects it is best compiled by the architect at concept stage. For complex projects, it may be coordinated and administered by a project manager.

A BPR includes aspects of a risk register widely used in construction, as well as elements of an energy risk register already used by some services engineers.[1] What differentiates a BPR from other risk registers is a holistic approach to capturing and tracking all the aspects of building design and performance identified through feedback studies that contribute to the performance gap: human factors, architectural elements, construction quality and operational issues.

Most importantly, a BPR provides the basis for contractual requirements in the tender documents which ensure that building performance is safeguarded during construction, commissioning, handover and operation. To close the performance gap and deliver zero carbon buildings, these guarantees must be embedded in the contract.

The BPR process creates shared incentives for the project team, including contractor and subcontractors, and encourages all team members to work towards the same goals. See figure 4.2.

The aim of the BPR is therefore to address the key risks identified in this book:

- Lack of contractual or legislative requirement to set, track or validate targets
- Unrealistic design assumptions in the absence of feedback from buildings in use
- Critical information relating to how these targets are intended to be delivered and by whom not being successfully communicated.

FIGURE 10.2
Building Performance Register (BPR): its objectives, what it will track, how it can help mitigate risk and, most importantly, assigning individuals to safeguard specific aspects of a project.

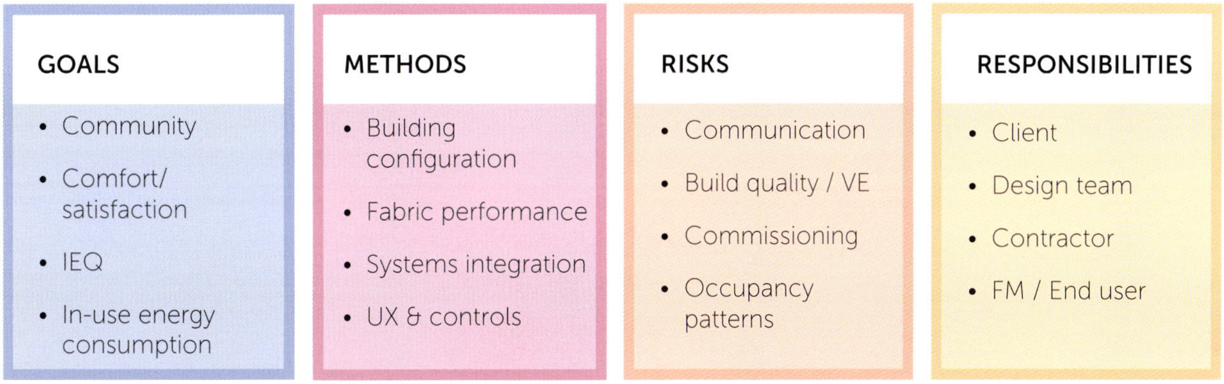

The case studies in chapter 9 demonstrate that where these issues are addressed, the design team stands a chance of delivering performance without major extra costs – see also note 10.01 for an example of how an architect's practice is embracing some aspects.

In summary, a BPR:

- itemises performance-related design assumptions and features at concept stage
- establishes a baseline that measured performance can be compared with
- safeguards material and product specifications so they can only be substituted by contractors and subcontractors if the alternative is shown not to affect energy and life-cycle costs (i.e. ringfencing)
- introduces accountability for building performance during construction
- provides the basis for enhanced building performance-focused commissioning, training, FM services and feedback
- helps ensure that metering is properly installed and commissioned so that post-completion data can be more easily collected
- establishes a framework for communicating building performance to occupants, including a building manual and logbook.

See also figure 10.4.

A BPR can be used to track key actions from briefing through to operation, and is presented in this book over five work stages, which are also a consolidation of the work stages in the RIBA 2020 Plan of Work[5] – see figure 10.3:

- ***Plan:*** Agree operational targets and create a BPR to track assumptions.
- ***Design:*** Develop design proposals that meet targets as set out in the BPR, refining and embedding them in drawings and specifications.
- ***Procure:*** Ringfence BPR targets and ensure quality control.
- ***Use:*** Verify BPR targets and undertake remedial action.
- ***Learn:*** Report on BPR targets and share lessons learned.

In summary, the BPR will tackle the targets as set out in table 10.1.

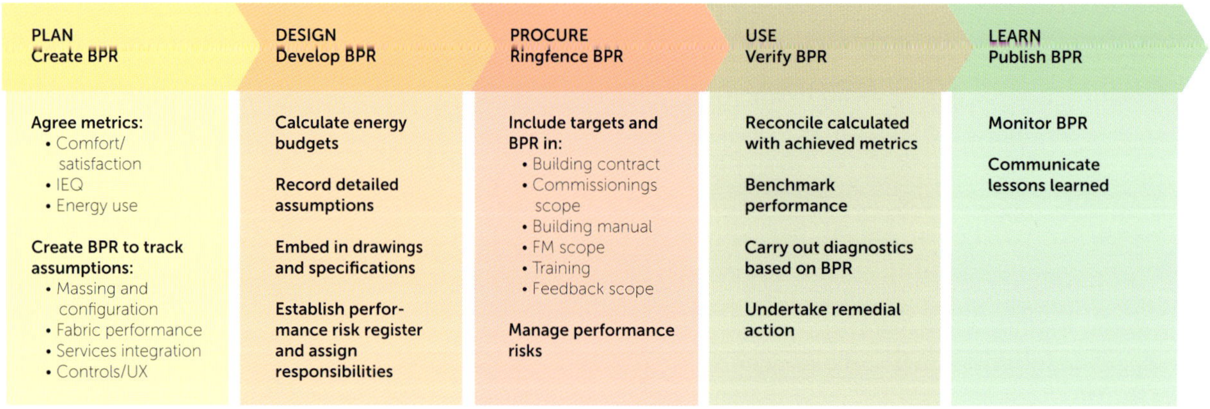

FIGURE 10.3
Diagram of a BPR, with the main activities to be undertaken at key project stages. See also table 10.1 and additional actions at the end of this chapter.

10.01: EXAMPLE

POE AT THE EVERYMAN THEATRE (CASE STUDY 1)[2]

Architects Haworth Tompkins have undertaken 'light-touch' POE studies cost-efficiently in-house to inform their design process, and began trialling this approach at the Everyman in 2019. Using the RIBA POE Primer[3] as guidance as well as recommendations from Fionn Stevenson's book *Housing Fit for Purpose: Performance, Feedback and Learning,*[4] the practice's scope includes:

- review of project information (drawings, initial sustainability strategy and aspirations, services strategy, specifications, etc.)
- site visits (layout alterations, changes of use, material weathering, performance of spaces, services and equipment)
- interviews with FM and other key staff
- environmental monitoring in key spaces (temperature, humidity, CO_2, VOCs, radon, etc.)
- analysis of energy data (from the BMS)
- BUS surveys for both permanent and transient users (training was undertaken to enable this)
- setting up an in-house methodology and reporting structure that can be adapted for other building types.

TABLE 10.1. QUALITATIVE AND QUANTITATIVE BUILDING PERFORMANCE TARGETS	
Occupant satisfaction (chapter 2)	Agree goals and benchmarks to evaluate occupant satisfaction, e.g. methodology, interviews/surveys or focus groups, but also building walk-throughs and observations, to capture: • occupant perception and satisfaction of indoor environmental quality (IEQ) • usability and maintenance of spaces and systems • space planning, design, wayfinding, access, productivity, perceived health and wellbeing.
Indoor Environmental Quality (IEQ) (chapter 2)	Set design criteria for comfort parameters: • thermal, including overheating • respiratory • visual • acoustic • usability (UX design).
Energy consumption (chapter 3)	Agree targets and scope to calculate, measure, validate and publish energy-use intensity data ($kWh/m^2/yr$): • by fuel (i.e. electric and non-electric energy used, generated, exported) • by end use (i.e. heating, cooling, ventilation, lighting, appliance loads, special equipment and other uses). Benchmarking is an essential part of the process.
Building configuration (chapter 5)	Agree approach to principles outlined in chapter 5, including: • context and massing, including overheating, pollution and noise, floor-to-floor heights • adjacency and zoning, including orientation, buffer spaces, thermal zoning, functional proximity, out-of-hours use.
Fabric performance (chapter 6)	Agree approach to meet fabric performance targets and ways to monitor and validate them, including: • airtightness • thermal resistance, including thermal bridging • vapour permeability • thermal mass • openings – windows, doors, rooflights and services, including fresh air, solar gain and glare, and acoustic conditions. Includes validation through performance monitoring such as occupant surveys, air permeability tests, thermal imaging, UX observations, hygroscopic and IEQ monitoring (e.g. temperature, relative humidity, light levels, dB monitoring – see also chapter 2).
Systems integration (chapter 7)	Agree approach to meet energy metrics and adopt Soft Landings process to manage risks for: • ventilation and cooling strategy, including air quality, energy savings, natural ventilation, MVHR • efficient heating and hot water • low-energy lighting • managing appliance loads • integration of low and zero carbon technologies. Energy in-use metrics together with IEQ and occupant satisfaction are proposed as a proxy for successful systems integration.
User experience (UX) (chapter 8)	Document design approach and quality-control process for controls and user experience, including: • control strategy, including manual and digital controls • configuring BMS, including zoning, occupancy, sensor location, set points, submetering, user experience, smart readiness • building manual • training.

Stage-by-stage actions for implementing the BPR are summarised in chapters 5 to 8. The following pages provide guidance on issues specific to procurement and contractual mechanisms for including and validating performance targets, in line with the RIBA Plan of Work stages.

10.02: DEFINITION

PRE-PROJECT SURVEY

A pre-project survey should assess the functionality and comfort conditions of end users' existing premises, including occupancy patterns and needs, maintenance regime and budget, and existing resource consumption. For projects where end users are not known, assumptions about occupant needs, flexibility and adaptability should be discussed with the client and recorded, along with adaptability and flexibility needs, to ensure long-term resilience.

BPR for planning and design (RIBA stages 0–2)

The scope and use of a BPR should be agreed at this early stage, together with the responsibility for updating and actioning items. Performance targets should be set with careful consideration of constraints and opportunities, and with the greatest possible engagement from the client to ensure success.[6]

At the briefing stage, the BPR should be a simple list of qualitative and quantitative energy, IEQ and other performance targets (see table 10.1 and chapter 2), and a strategy for meeting them. To deliver truly zero carbon buildings, the targets may also include indicators for embodied energy and water consumption, as set out in the RIBA 2030 Climate Challenge.[7] The EU Level(s) programme offers a further range of metrics for climate change resilience and life-cycle costing.[8]

For setting energy targets at the briefing stage, performance metrics can be based on a combination of industry benchmarks for similar buildings and experience (see chapter 3). Whether an existing benchmark is used, or a project-specific benchmark calculated or modelled, should be agreed now. At this early stage, total operational energy consumption 'by fuel' is usually adequate. This could include a breakdown of all grid electricity used, renewable electricity generated and exported, fuel consumption and any heat to be exported – or just a total energy-use figure. Targets are assumed to be normalised for weather.

Interviews with different stakeholders and a pre-project survey can also provide valuable input and a baseline for comparison once the project is complete – see note 10.02. This is to capture client and end-user priorities, expected usage patterns and design assumptions. Where a pre-project survey is carried out, the same survey methods should be used post-project completion. The method and number of post-completion surveys should be agreed upfront.

How a project may contribute to social and community welfare is also important to record, and can be captured from early meetings.

BPR for procurement at early stages: design team scope of services

To help clients assess the capability of design team members to work towards in-use quantitative and qualitative indicators, measurable building performance targets should be included in the project pre-qualification questionnaire – see table 10.1.

Independent validation of contractual performance targets should be considered, especially if design and build (D&B) procurement is likely, or where the design team lacks feedback expertise in-house. If an independent building performance consultant is appointed, then providing information to them should be incorporated into the project team scope. For design and build contracts, where design teams are novated to the contractor, reporting on indicators and updating the BPR is fully assigned to the contractor, with adequate client-side oversight.

As part of the design team appointment, any additional scope of works should be agreed for each consultant. Typical additional services include:

- creating and updating the BPR at key project stages
- indoor air quality, lighting and acoustics analysis
- overheating analysis to include realistic occupancy scenarios and equipment loads
- calculating, updating and sharing operational energy end-use budgets with the project team (see CIBSE TM54[9])
- drawing up design team and contractor incentives for meeting project zero carbon goals
- managing the 'performance contract' for energy and IEQ, including tracking that the BPR items are implemented during construction
- reconciliation of control strategy with as-built building controls and meters
- gathering feedback and comparing results with project goals
- preparation of a report to communicate lessons learned.

BPR for design development (RIBA stages 3–4)

Design development is critical for translating high-level targets established at the briefing stage into realistic proposals. At this stage BPR can take the form of a written report detailing the design and system strategy as well as a building's specific features determining energy performance. The architectural and spatial design features that directly impact building performance, such as floor-to-floor heights, the configuration of openings, and fabric build-up (see chapters 5 and 6), must be documented in the BPR so they can be tracked through subsequent stages. The report should be accompanied by a schedule listing such features as well as the assumptions about use and maintenance.

Energy targets and benchmarks set at the briefing stage are translated into energy end-use 'budgets' in kWh/m^2/yr – see 'frameworks for energy prediction' in note 10.03. This includes heating, cooling, ventilation, lighting, appliance loads and any special equipment. Technical building systems, zoning and controls should also be considered (see chapters 6 to 8). Architects should undertake initial massing and energy modelling and heat loss and overheating studies in-house, especially when engineers are not appointed at the early stages – see table 10.1 and note 10.04.

Sensitivity analyses should also test realistic operating scenarios, including different zones, hours of use and occupant density. Any high energy-intensity functions, such as server rooms, cafeterias, laboratories and so on, should be included for in the design at this point. The flexibility of the building to accommodate different functions and related variations in occupancy with minimal changes to fabric and systems is important for long-term resilience.

10.03. EXAMPLE

FRAMEWORKS FOR ENERGY PREDICTION

In the UK, CIBSE TM54 for buildings offers a methodology for design calculations that accounts for a wide range of actual occupancy and equipment-use scenarios. In the US and elsewhere, ASHRAE 90.1[10] (non-domestic) & 90.2[11] (domestic buildings) provide relevant guidance. In addition, Passivhaus, Soft Landings and Design for Performance all offer step-by-step guidance on predictions. See also chapter 3.

10.04. EXAMPLE

DIGITAL TRACKING OF BUILDING PERFORMANCE

EU Building Passports[12] are envisaged as a digital online record of building properties that determine building performance. In the UK, EPC data files and DEC data entry fields provide a useful template for recording minimum building and system properties; these can be complemented by information on actual hours of operation, occupancy, appliance loads and special functions.

10.05. EXAMPLE

DIGITAL TRACKING OF TARGETS THROUGH BIM

Building information modelling (BIM) can provide an opportunity to digitise the tracking of factors contributing to performance. For projects looking to go beyond implementing BIM Level 2,[13] tagging and change control of building features that relate to energy efficiency may lead to more efficient management of building performance risks.

Simultaneous updating of energy models can inform decision-making about different design scenarios, and could even be tracked through building information modelling (BIM) – see note 10.05. Modelling input assumptions should be documented and shared in the BPR to add an extra level of rigour to energy calculations. Once agreed, detailed energy end-use budgets should be logged in the BPR along with any relevant strategies for meeting them. For example, if the planting of deciduous trees for summertime shading is proposed, this would need to be included in the register, because it makes a critical contribution to limiting summertime overheating.

The selection of technical systems and the zoning of controls must all be determined at this stage, carefully anticipating how they will be used based on maintenance expertise and budgets. The BPR should include a non-technical version of the BMS control strategy. During detail design and technical stages, risks related to value engineering should be added relating to each item on the BPR so that these can be managed.

The BPR should ideally include a summary of the life-cycle costing report so that long-term energy and maintenance impacts are acknowledged from the beginning.

Finally, as part of recording key factors for building performance, measures that aim to improve health, wellbeing and community value should be included. For example if a company is upgrading premises to improve employee satisfaction, then describing the aspects of design that respond to employees' needs should be included.

BPR for procurement (RIBA stages 5–6)

At procurement stage, it is very useful to ask a contractor, and any named subcontractors, to demonstrate their knowledge and experience in delivering operational performance targets. Once the building contract is let, the BPR turns into a risk register, which means that major risks of change against each item must be evaluated and actioned. While it is not a guarantee of construction quality, the BPR will raise a project team's awareness of the issues and provide a means to hold a contractor accountable once feedback is available.

Performance targets must be incorporated into the contractor's preliminaries and linked to financial incentives if they are to be contractually binding. A bonus is recommended for meeting operational targets (or a penalty when not met), and the time of release can be by mutual agreement. If the building performs as expected, the contractor can complete their involvement at the end of seasonal commissioning, i.e. by the end of year one of operation. However for most buildings, a three-year period is a realistic time frame for the contractor to demonstrate that project performance indicators have been met.

A performance bonus and the time frame for releasing it should be agreed for meeting energy and IEQ. The main advantage of this is that issues are identified and fixed during the defects liability period, saving the client time and funds. Monthly performance data drops are required for this to work, and to facilitate the reconciliation of design versus as-built performance.

Incorporate BPR into the construction stage

During construction, a contractually binding BPR can mitigate the three areas that pose the greatest risk to building performance:

- Value engineering: the replacement or omission of specified items or processes with products or materials that are of inferior specification and/or do not meet the performance criteria set out at design stage
- Construction quality issues affecting fabric performance
- M&E installation and commissioning.

If the main contractor, and in turn individual subcontractors, have a clear responsibility for delivering all the items, including the BPR targets, inappropriate value-engineering can be minimised – see note 10.06

The in-use performance indicators should remain unchanged from tender to handover stages, unless changes to the BPR are explicitly agreed with the client and design team. Contractors can replace items as long as performance outcomes and the impacts of any changes on the overall design remain unaffected. This means not just energy targets, but also comfort targets and life-cycle costs, and how they are achieved. For example, if a vapour-permeable element or material is substituted for another material with the same thermal performance but less permeability, this may lead to worse IEQ, which should be flagged by the contractor.

Thermal imaging and airtightness tests should be undertaken at the construction stage. If substantial moving out of occupants took place, an occupant survey can offer invaluable insights into the priorities of occupants and inform the handover process.

Incorporate BPR into building commissioning

Given that all performance-related features are in a single accessible document, the BPR streamlines the process of enhanced commissioning, ideally through an independent commissioning agent – see note 10.07. The BPR helps inform the creation of a building manual, and is the starting point for subsequent building performance evaluation (BPE).

BPR in-use and for learning (RIBA stage 7)

Building operation starts at the point of handover, and this is a crucial time in terms of performance. In the operational phase, the BPR becomes a template for BPE, which usually starts with the reconciliation and analysis of energy data to highlight any performance anomalies. Therefore, ideally from the first day of operation, the contractor should start collecting energy data and sharing it with the client, the project team and whoever

10.06: EXAMPLE

MINIMISING VALUE-ENGINEERING OF PERFORMANCE TARGETS

At Keynsham Civic Centre (case study 6), value-engineering of critical items was discouraged due to the contractual requirement for the contractor to demonstrate (with an updated energy model) that any changes would not affect performance targets listed in the BPR.

10.07: DEFINITION

ENHANCED COMMISSIONING

This is the process of ensuring that all equipment has been installed, calibrated, linked up and profiled, and is communicating correctly as prescribed by the project's controls strategy (see chapter 8) and BPR performance targets. It also includes seasonal commissioning, i.e. validation in all thermal control seasons. Occupant satisfaction interviews also form part of this process. The Soft Landings programme provides guidance for actions that support quality control related to the installation and commissioning of technical problems.

10.08: DEFINITION

PERFORMANCE VALIDATION

Validation of performance means that the energy performance baseline is updated to reflect the as-built building, and actual operating conditions are compared with measured data for both energy and IEQ parameters. If the results are within an agreed range, then the building is assumed to be operating as intended. Sometimes it takes several years for a building to reach this state, as in the case of case study 6, where three years after handover the building reached its performance equilibrium.

is responsible for interpreting the feedback data. At this stage, the building energy model may need recalculating to match the as-built project properties and actual operating conditions.

The scope and deliverables should be defined for BPE, which can still be procured at this stage – the Soft Landings aftercare framework offers helpful guidance on making this happen. The first year of operation is part of the defects liability period, so information about building performance is particularly sensitive, because it may expose underlying defects. This is why agreeing responsibilities for performance and feedback processes upfront is preferable, including clarifying the responsibility and authority to act on feedback findings. It is important to include measures to cover the risk of escalating costs if difficulties arise with collecting data and project documentation. It may also be necessary to advise insurers and agree a separate contract for feedback processes.

FIGURE 10.4

Diagram summarising the BPE processes and a graduated response approach with regard to carrying out a comparison of building energy use throughout different project stages.

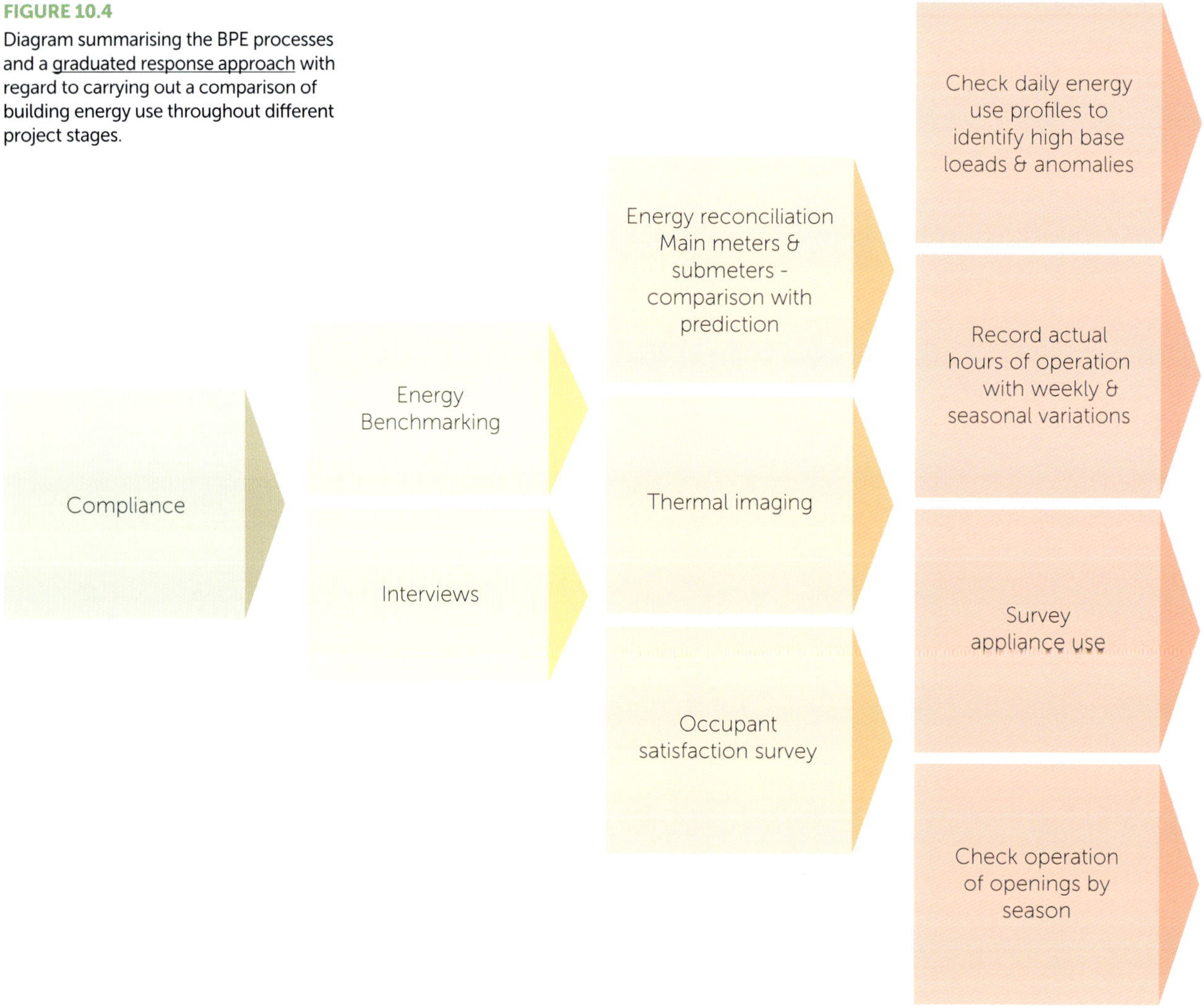

If performance targets and feedback have been contractually agreed in the BPR, it is in the contractor's interest to examine the building to uncover and rectify issues to demonstrate that targets are met, or exceeded. In this case, the client can use the BPR to evaluate if the contractor has delivered a building that performs as intended.

The BPR also provides a helpful basis for the building FM scope, and can underpin a performance contract for building management and frame any training or re-commissioning where necessary.

Finally, when it comes to sharing lessons learned on a project with the design team, within consultant practices or industry-wide, the BPR is an invaluable source of information for understanding and communicating building performance targets and how they were met. Successful approaches can then be replicated in future projects.

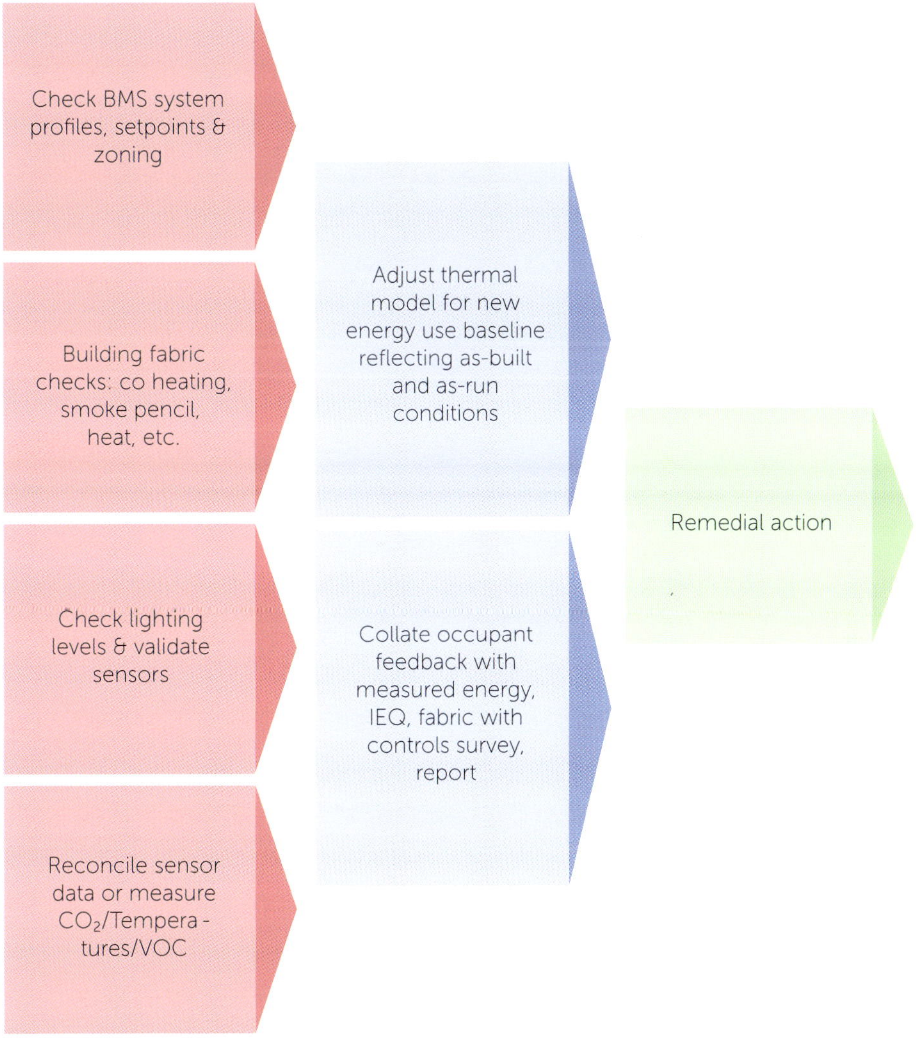

Gathering feedback

It is important to start the feedback process while the design team and contractors are still engaged with the project, to be able to source information about design intent and to act on findings. Gathering feedback involves:

- comparing design with as-built to control quality
- validating performance by updating the BIM to include as-built and operating conditions, including IEQ, and comparing with measured – see note 10.08
- comparing with qualitative data collected from occupant surveys and observations from site visits
- undertaking remedial actions, and continuing the above data collection after remedial measures
- sharing lessons learned internally within the design team, and publicly.

ACTIONS	PLAN	DESIGN
	• Check with insurers, and if necessary set up a separate contract for feedback processes. • Draw on feedback experience and evidence published by others at the concept stage to avoid perpetuating mistakes. • Conduct a pre-project survey with existing users (if applicable, see note 10.02) and with the FM and other stakeholders.	• Compare estimated design stage performance with chosen benchmarks. • Establish the BPR as a written report detailing the design and system strategy as well as a building's specific features determining energy performance. • Document the architectural and spatial design features that directly impact building performance in the BPR. • Record the way in which technical systems rely the spatial and material configuration of the building • Incorporate in BPR assumptions behind technical systems and the controls strategy, including key aspects of UX design

Contractors can only be obliged to facilitate a BPE if this is incorporated into the project contract. At the time of writing, there are no contractual industry standards for feedback processes, but certification schemes such as Passivhaus, Design for Performance and several of the case studies described in chapter 9 provide precedents.

Feedback can be carried out by an independent agent, in which case a contractor must agree to BPE during the first year of operation within the defects liability period. Under a performance contract, the contractor is also engaged to remedy faults.

Early client commitment to building performance is a major factor in success. Early client engagement of an expert member of the design team or an independent building performance specialist can optimise the feedback process. This is especially helpful during construction and post-handover to oversee the contractor's delivery of the BPR and to interpret feedback outputs and data collection.

EARLY CLIENT ENGAGEMENT OF AN EXPERT MEMBER OF THE DESIGN TEAM OR AN INDEPENDENT BUILDING PERFORMANCE SPECIALIST CAN OPTIMISE THE FEEDBACK PROCESS.

PROCURE

- Ask the (sub)contractor to demonstrate their knowledge and experience in delivering operational performance targets.
- Incorporate into the contractor's contract financial incentives for meeting performance targets, along with the need to recalculate impact on performance aspects (energy, comfort, etc.) if changes are made to demonstrate that the change did not affect project targets.
- Ensure that that all equipment has been installed, calibrated, linked up and so on as prescribed by the project's controls strategy.
- Create a building manual and subsequent BPE, including documentation on use and maintenance including building configuration, build-up, services and controls.

USE

- Clarify lines of responsibility for data collection and remedial action.
- Schedule seasonal commissioning and occupant satisfaction interviews.
- Ensure that controls of technical systems have been optimised, commissioned and issues rectified.
- Compare estimated design stage performance with in-use performance and benchmarks.

LEARN

- Include reflections and a clear list of lessons learned, and agree key messages with all of the stakeholders in the design team prior to wider sharing.
- Obtain permission from the client for wide sharing and publication of lessons learned; consider making these anonymous and more generic for wider dissemination.
- Discuss with the client how 'lessons learned' can be used to support the FM team.

EVERYMAN
PLAYHOUSE

Conclusion

Energy, People, Buildings is a call for designing with evidence – gathering feedback from how buildings actually work – to chart a path towards zero carbon buildings. Over the next decade, feedback must become a part of every architect's toolbox to ensure that buildings perform as intended to meet the imperative of climate change. Design based on intuition is no longer good enough.

Building performance feedback is a proven approach to ascertain what works and what doesn't on the route to zero carbon. Targeting and tracking the right metrics throughout design and construction – and in use – is a collaborative cross-disciplinary process that can help overhaul design and procurement to deliver zero carbon buildings that work and that close the performance gap.

To meet 2030 climate targets, every energy end use in a building must be scrutinised and challenged by every member of the project team. This requires energy literacy, as well as a rudimentary understanding of the subjects tackled in the different chapters of this book: comfort, passive design, fabric performance, appropriate technical systems and user-friendly building controls.

Metrics can be transformative, not just for energy but for comfort and quality, too. Traditionally, designers have not been deeply versed in the technicalities of building performance, yet a new generation of architects and engineers is emerging who are energy and climate literate, who understand passive design and fabric performance, and who can use energy models as a decision-making tool in their designs.

FIGURE 11.1: OPPOSITE
Everyman Theatre, Liverpool, Haworth Tompkins.

This book demystifies and distils the key ingredients of building performance evaluation in the hope that feedback will increasingly become part of designers' core skill set. The key messages that emerge from this book chime with those identified by the Rocky Mountain Institute (case study 4) as lessons learned from its Colorado headquarters:

- Integrated design is crucial to deliver cost-effective zero carbon buildings.
- Understanding people's behaviour and getting their buy-in is essential.
- Commissioning and monitoring are critical.
- There is a need for user-friendly meters and controls.
- Performance contracting helps manage risk.

The concept of a Building Performance Register is proposed as a way to track the complex process of sustainable design throughout the life of a project. It is the one document and process that encapsulates and safeguards all the performance aspects of a building and promotes a shared understanding across the entire project team. Many of the lessons learned from the non-domestic case studies and building performance evaluations featured throughout this book also apply to housing and other building types.

The climate crisis means that any new buildings which architects design now face an uncertain and potentially challenging future. How these buildings will perform in real life depends largely on the feedback processes designers use to ensure that what is built meets their targets and intentions. The existing building stock must urgently be upgraded to higher standards, and feedback processes can inform the targets and ambitions of the retrofit agenda. Making explicit what people value in existing buildings can help speed the transformation that is so urgently needed. Enormous challenges lie ahead, yet the 2020 pandemic has revealed the global potential for drastic collective action. Similar mobilisation must be applied to the far bigger climatic challenge ahead.

The need to design with a light footprint on the planet is not just about thermal performance and energy metrics. It's a call to challenge conventional ideas of beauty and comfort and invent a new architectural aesthetic which can steer the profession's response to climate change. A thorough grounding in evidence can, and must, show the way.

FIGURE. 11.2 CASE STUDIES ENERGY USE INTENSITY (EUI)

Figure 11.2 shows the annual measured energy use intensity (EUI) for all the Chapter 9 case studies compared with the RIBA 2030 Climate Challenge benchmarks. The graph shows total EUI for heat (red) and electricity (blue) offset by any renewable energy (green) generated and used on site.

The four projects with the lowest EUI were all built to Passivhaus or equivalent standards, while Keynsham Civic Centre and the Rocky Mountain Institute were built under a performance contract.

Both UK office buildings, Keynsham Civic Centre and Morelands Rooftop, are intensively used with extensive out of hours operations and large IT loads which are reflected in higher per m2 consumption. The Everyman Theatre is equipped with professional audiovisual equipment but its consumption is still in line with the CIBSE good practice benchmark for this building type.

Although designed almost a decade ago, the Enterprise Centre and the Rocky Mountain Institute exceed the RIBA 2030 targets even before onsite renewable energy generation is taken into account. In addition, Keynsham Civic Centre and Frederiksbjerg School are within close range of the 2030 target.

What all these projects share is a feedback process which informed their design, construction and operation. Clearly both Passivhaus certification and performance contracting offer tangible benefits in getting to net zero.

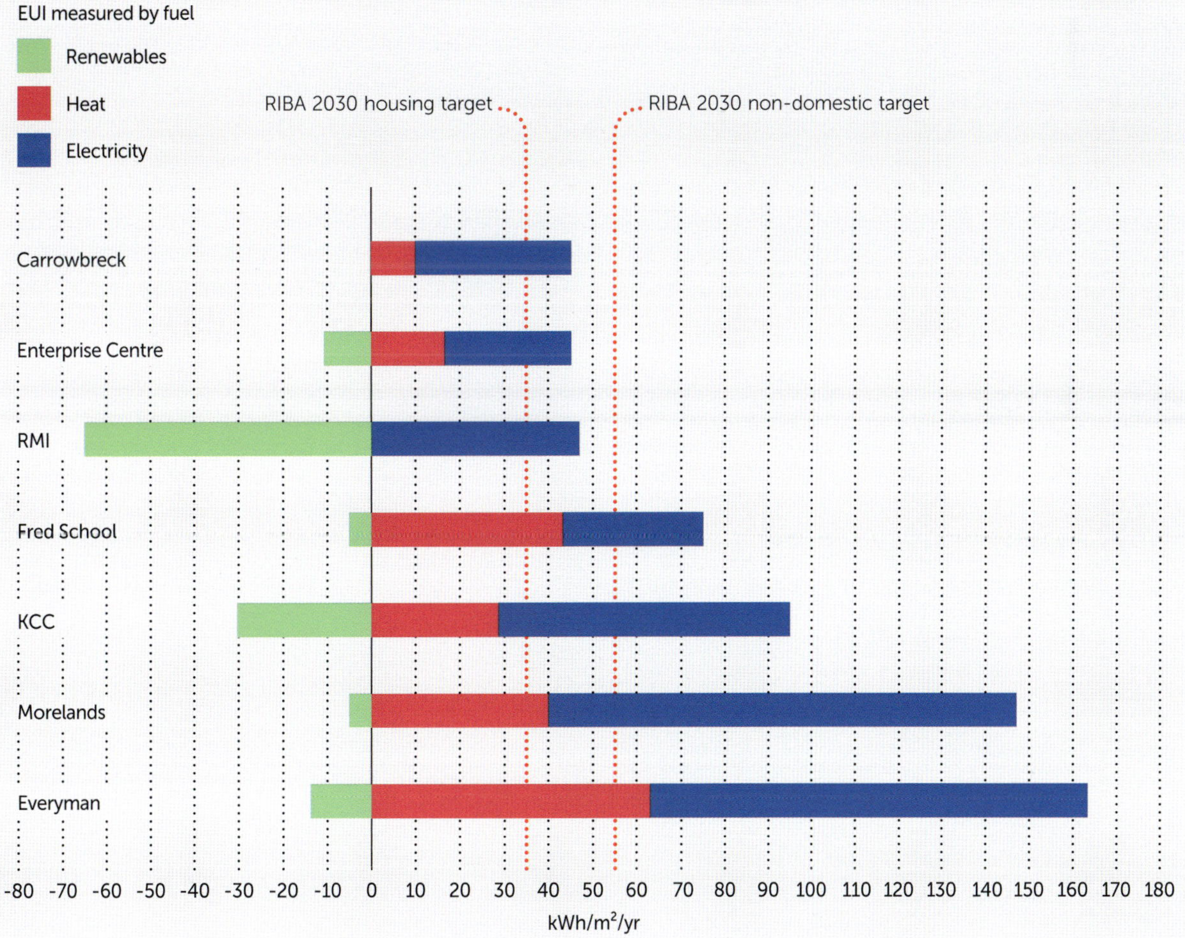

Glossary

Access/usability of controls
Well-planned control and positioning of non-fixed systems is important: switches must be easily accessible and unobstructed.

Adaptive comfort – see page 21

Air-handling units (AHUs) – see chapter 7
Air-handling units are (large) appliances that filter and condition external air, and distribute it within a building.

Airtightness[1] – $m^3/(hr.m^2)$ or ach^{-1}
Measured as cubic metres (volume) of air exchange per square metre of building envelope per hour, under a 50 Pa pressurisation test. You can also use ach^{-1} (air changes per hour at 50 Pa), which is more commonly used elsewhere in Europe, for example in the Passivhaus standard. Consult local regulations and standards to assess the required airtightness targets.

Auditory or acoustic comfort[2] – see page 24

Base building performance[3] – see page 34
Performance based on the energy used for common parts of building and shared services. Part of NABERS and Soft Landings.

Benchmarking of energy performance[4] – see page 33

Biophilia[5] and biophilic design – see page 22
An architecture that aims to connect people with the natural world through natural ventilation, and access to light, sun, external views, etc.

Building (energy use) diagnostics[6]
Techniques to evaluate the performance of a building or one of its systems. These diagnostics are usually carried out by a mechanical engineer with building performance evaluation training.

Building management/facilities management[7]
Services for the management of technical systems and controls, the maintenance regime, basic repairs, and arranging expert repairs. In larger buildings this is referred to as facilities management.

Building management system (BMS)[8] – see page 98

Building performance evaluation (BPE) – see page 39

BPE programme[9]
A four-year Innovate UK programme that analysed domestic and non-domestic buildings to evidence the causes of the performance gap.

Building Performance Register – see page 173

Building Use Studies (BUS) methodology[10]
A standardised questionnaire used to benchmark user satisfaction on a wide range of factors relating to design, thermal and acoustic comfort, air quality, spatial functionality and management.

CarbonBuzz[11] – see page 33
An online platform used to record and share building energy use in operation (based on CIBSE benchmarks), and compare it to the design expectations.

CIBSE[12]
Chartered Institution of Building Services Engineers. It publishes Technical Memoranda (TM) that are often used to provide guidance on environmental systems and technologies.

Circadian rhythm[13]
Physiological, mental and physical changes that follow the daily cycle of day and night and affect sleep patterns and mental health.

Carbon/CO_2 intensity factors[14] – see page 29

Coefficient of performance (COP)[15]
Ratio of heating or cooling of a system according to the energy input required. A higher coefficient implies greater efficiency.

Combined heat and power (CHP) – see page 88

Commissioning
Making installations ready for service, implying that all installations are tested, quality controlled, and comply with all regulations and all design specifications when they are commissioned. Commissioning of energy systems requires testing during three thermal control seasons: summer, winter and autumn/spring.

Compliance calculations/certification model
These are calculations or simulation models used for regulatory approval. They typically do not include occupant behaviour or unregulated loads, and often do not include detailed site information. Typically these models are based on **steady-state models.**

Deadband[16]
A gap between the set-point values in one or multiple systems where nothing happens. For instance, between the set-points for the heating to switch off (at 21°C) and for the

cooling to switch on (at 23°C), there is a 2°C 'deadband' where neither heating nor cooling occurs. Without an appropriate deadband, automated systems can behave erratically or even at the same time towards opposing goals (e.g. heating and cooling simultaneously).

Demand-side management[17] (DSM)
The ability to automatically shift a building's demand load to draw electricity from the grid during off-peak times, to minimise costs and balance the grid load. Additionally, DSM stipulates that, if any spare electricity is produced on-site, it is fed back to the grid at peak times.

Dependencies
Conditions used to regulate an automated building management system. The greater the sophistication of the system, the more articulated the dependencies can be. For instance, artificial lighting can be controlled both by the presence of occupants in a room and by the intensity of daylight: if the classroom is occupied and daylight levels reach 250 lux, the lighting is dimmed to half its output. These conditions must be purposely embedded in the management system based on evidence and planned outcomes.

Display Energy Certification (DEC) (UK only)[18] – see page 34

Dynamic model[19] – see page 41
The simulation of energy use with changing internal (occupancy, out-of-hours, etc.) and external (weather, season, daylight, night cooling, etc.) conditions over time. Software such as IES or EnergyPlus offer dynamic energy modelling of buildings.

Earth tubes – see page 87

Embodied carbon[20]
The sum total of CO_2 emissions associated with the manufacturing, transport, use, decommissioning and disposal of any object. Not usually regulated for.

Energy Star – see page 33

Enhanced commissioning – see page 179

Energy Performance Certification (EPC)[21] (England and Wales)
A domestic energy-efficiency certification system based on a non-invasive survey of the property.

EPBD[22]
Energy Performance of Buildings Directive – part of the EU legal framework for a decarbonised building stock by 2050.

Feedback – see page 39

Feedback loops
A feedback loop occurs when the system's output (for instance energy consumption readings) is used as input for future operations (e.g. the system is adjusted to better match the expected performance).

Forgiveness factor[23] – see page 19

Form factor[24]
Building compactness is the relationship between the exposed building surface area (A, m^2) and the building's volume (V, m^3). For the Passivhaus standard, instead of building volume, the internal treated floor area is used for the form factor.

Glare[25]
Visual discomfort caused by excessive luminance or contrast.

Graduated response approach[26]
Building evaluation carried out starting from a more general evaluation, and moving towards greater detail as needed or warranted.

Heat pumps[27]
A device transferring heat from a 'source' to a 'sink' against the normal flow of heat (the source is colder than the sink). The heat source can be air, water or the ground.

Hygroscopicity – see vapour permeability

Indoor environmental quality (IEQ) – see page 17

Indoor air quality (IAQ)/respiratory comfort[28] – see also page 17
Refers to the air quality inside or in the close vicinity of a building, affecting the health and comfort of building occupants.

Interstitial condensation[29]
Occurs when humid air diffuses into a vapour-permeable material and reaches the 'dew point' within the material, condensing into water in the construction build-up.

Key performance indicators (KPIs)[30]
Key performance indicators are measurable milestones agreed by the client and design team. Reporting is usually undertaken at different stages of the project to ensure that the objectives are validated at design, commissioning and in-use stage.

Level(s) standard[31]
A voluntary sustainability framework to improve the performance of buildings, developed by the EU and intended as a common approach within the EU.

Light transmittance (g-value)[32]
The amount of light that can pass through a material without being absorbed or reflected.

Living Building Challenge[33]
A design philosophy and a building certification tool centred around seven performance categories based on actual rather than design performance; at least 12 consecutive months of building operation are required post-construction. The building's energy needs must be met 105% by on-site renewables, and this excludes on-site combustion.

Meter reconciliation – see pages 30 and 102

Mixed-mode ventilation[34]
A ventilation approach that combines natural ventilation with mechanical systems (of any type), with the goal of minimising mechanical cooling loads.

NABERS (National Australian Built Environment Rating System)[35] **– see page 35**

Net zero energy, net zero carbon, zero carbon, carbon neutral, nearly zero – see page 29

Non-fixed equipment
Any appliance, of any kind, that is not part of the building design or systems.

Occupancy patterns[36]
Occupant density, hours of use, intermittent occupancy and variations of occupant density in different locations in the layout (i.e. the same number of people in the building can spread out or gather for talks, parties, lectures or meals in larger groups and leave other areas sparsely occupied).

Occupant density profile[37]
The number of occupants in a building, and their use of the building and its systems.

Occupant satisfaction
A measure (normally qualitative) of the level of satisfaction users have with the building and the services it provides.

Operating conditions[38]
Includes **occupancy patterns** as well as **building management.**

Operational energy performance and operational emissions – see also page 34
The energy (or carbon emission equivalents) used 'during the building lifetime' (in-use) for heating, cooling, lighting and appliances. CIBSE TM54 can be used to evaluate operational energy performance at the design stage.

Passivhaus standard – see page 35

Performance – see page 4

Performance gap – see pages 4 and 9

Performance contract/contracting of building management
A building management contract drawn up with financial incentives to meet specific targets.

Performance validation/validation of performance/ validation of in-use performance – see page 180
Assessment of the 'in-use' energy performance of a building to validate the 'as-designed' performance expectations.

Predicted mean vote (PMV)[39]
PMV expresses thermal satisfaction based on the average satisfaction of a large group of individuals, using the ASHRAE thermal sensation scale with seven values from hot (+3) to cold (-3).

Post-occupancy evaluation (POE) – see page 4

Power rating[40]
The number of machines installed (density) and the efficiency of the equipment, as well as how long they are used for. System maintenance and correct design and installation affects the machines' performance and thus the actual rating.

Primary energy use
The total energy demand of a building. To report energy use as a single total figure in kWh, thermal and electrical energy must be converted to a primary energy use.

PROBE studies[41]
Post Occupancy Review of Building Engineering Studies, a set of CIBSE cases studies describing practical evaluation for different buildings.

Productivity gains
The productivity increase (i.e. the effectiveness of effort in doing work and associated savings) due to the increased comfort and efficiency in a building with high indoor environmental quality (IEQ).

Regulated and unregulated energy[42]
Regulated energy is the energy consumed by building systems subject to building regulations. Unregulated energy is the energy consumed by non-fixed equipment, and is not directly subject to building regulations.

Sensitivity analysis
A simulation analysis used to determine the contribution of each design variable to the energy performance (or other agreed metric) of the building. Run iteratively, modifying different parameters, it can identify better approaches to energy performance.

Set points
Specific values obtained from sensors used by the BMS to alter the behaviour of a specific system, typically referred to for temperature settings.

Soft Landings[43]
A framework that aims to improve operational building performance, occupant satisfaction and client expectation, and does this by supporting building procurement from the early stages to building completion and occupation.

Steady-state model[44]
The simulation of energy use based on the assumption that indoor and outdoor conditions are stable and unchanging over time. Applications such as PHPP, SAP and SBEM support steady-state building models.

Thermal alliesthesia – **see page 21**

Thermal break[45]
An element with low thermal conductivity used to prevent the creation of a thermal bridge.

Thermal bridging[46] – **see page 68**

Thermal comfort[47] – **see page 21**

Thermal mass[48] – **see page 68**

Thermal transmittance (U-value) and thermal resistance (R-value)[49]
The rate of heat transfer through a structure in W/m^2K. The lower the U-value the higher the insulating property. The inverse of the heat transfer through an element is the thermal resistance, thus the higher the R-value the higher the insulation properties.

TM54[50]
CIBSE Technical Memorandum 54: Evaluating Operational Energy Performance of Buildings at the Design Stage.

TM55[51]
CIBSE Technical Memorandum 55: Design for Future Climate: Case Studies

Schedule
In a BMS, the (standard) lists of tasks and times for the operation of building services.

Urban heat island effect[52]
The observed higher temperatures in urban areas compared to the surrounding rural areas, caused by human activities and thermal mass of buildings.

User experience (UX) design
Design for the user's experience, typically referring to a digital system and the user interface.

Vapour permeability[53]
A material's ability to permit moisture (water vapour) to pass through it. Hygroscopicity refers to a material's ability to absorb and release water vapour without reducing its thermal performance.

Variable-speed fan drive and variable air volume (VAV)[54]
Variable-speed fans are devices that can vary the speed of a fixed-speed motor, used in a VAV (a specific kind of HVAC) unit to vary the air flow at a constant temperature.

Ventilation: background ventilation, purge ventilation, mechanical extract ventilation (MEV) and mechanical ventilation with heat recovery (MVHR)[55]
Background ventilation can be provided by continuous passive (trickle vents) or active (MEV or MVHR) air flow. Purge ventilation is ventilation that replaces hot indoor air with cooler outdoor air, typically needed in summer.

Visual comfort – **see page 23**

Whole-life carbon emissions[56]
All the emissions arising during the whole life of a building, including those from building operation embodied carbon, ideally accounting for future reuse and recycling of any part.

Whole-life costing/life-cycle costing[57]
The financial implications of materials, products and systems over the building's intended life span, including maintenance and replacement in the building's total capital and operational costs over the building life span.

Further reading
(also on www.energypeoplebuildings.com)

Books

Clark, D. (2013),*What Colour is Your Building? Measuring and Reducing the Energy and Carbon Footprint of Buildings*, RIBA Publishing

De Wilde, P. (2018), *Building Performance Analysis*, Wiley-Blackwell

Dollard, T. (2017), *Designed to Perform: An Illustrated Guide to Delivering Energy Efficient Homes*, RIBA Publishing

Mumovic, D. (2018), *A Handbook of Sustainable Building Design and Engineering: An Integrated Approach to Energy, Health and Operational Performance*, Routledge

Partington, R. and S. Bradbury (2017), *Better Buildings: Learning from Buildings in Use*, RIBA Publishing

Penoyre & Prasad (2014), *Retrofit for Purpose: Low Energy Renewal of Non-Domestic Buildings*, RIBA Publishing

Stevenson, F. (2019), *Housing Fit For Purpose: Performance, Feedback and Learning*, RIBA Publishing

Reports and guidance

Better Building Partnership (2019), *Design for Performance: A New Approach to Delivering Energy Efficient Offices in the UK*, https://www.betterbuildingspartnership.co.uk/sites/default/files/media/attachment/BBP_Design%20for%20Performance_A%20new%20approach%20to%20deliver%20energy%20efficient%20offices_0.pdf (accessed 14 August 2020)

Building Services Research and Information Association (2015), *Performance Evaluation in Non-Domestic Buildings (BG 63/2015)*, https://www.bsria.com/uk/product/XBYYwB/building_performance_evaluation_in_non_domestic_buildings_bg_632015_a15d25e1/ (accessed 14 August 2020)

Building Services Research and Information Association, Soft Landings guides and framework, https://www.bsria.com/uk/consultancy/project-improvement/soft-landings/guides/ (accessed 14 August 2020)

Chartered Institute of Building Services Engineers (2006), *TM22: Energy Assessment and Reporting Methodology*

Chartered Institute of Building Services Engineers (2013), *TM54: Evaluating Operational Energy Performance of Buildings*

Chartered Institute of Building Services Engineers (not yet published), *TM61: An integrated framework for building performance*

Gething, B. (2010), Design for Future Climate, *Opportunities for Adaptation in the Built Environment*, https://www.arcc-network.org.uk/wp-content/D4FC/01_Design-for-Future-Climate-Bill-Gething-report.pdf (accessed 14 August 2020)

Danish Association of Architectural Firms (2020), *Architecture Creates Value*, https://www.danskeark.dk/sites/default/files/2017-06/Architecture_creates_value.pdf (accessed 14 August 2020)

Palmer, J. et al. (2016), *Building Performance Evaluation Programme: Findings from Non-Domestic Projects – Getting the Best from Buildings*, Innovate UK, https://assets.publishing.service.gov.uk/government/uploads/system/uploads/attachment_data/file/497761/Non-Domestic_Building_performance_full_report_2016.pdf (accessed 14 August 2020)

Palmer, J. et al, (2016), *Building Performance Evaluation Programme: Findings from domestic projects – Making Reality Match Design*, Innovate UK, https://assets.publishing.service.gov.uk/government/uploads/system/uploads/attachment_data/file/497758/Domestic_Building_Performance_full_report_2016.pdf (accessed 14 August 2020)

RIBA (2018), *Ethics and Sustainable Development Commission: Final Report*, https://www.architecture.com/knowledge-and-resources/resources-landing-page/ribas-ethics-and-sustainable-development-commission-final-report (accessed 14 August 2020)

RIBA (2020), *Plan of Work 2020*, https://www.architecture.com/knowledge-and-resources/resources-landing-page/riba-plan-of-work (accessed 14 August 2020)

RIBA, Post Occupancy Evaluation Guidance, www.architecture.com/knowledge-and-resources/resources-landing-page/post-occupancy-evaluation (accessed 14 August 2020)

UK Green Building Council, Practical How-To Guide: How To: Execute a High Impact Post Occupancy Evaluation, https://www.ukgbc.org/sites/default/files/How%20to%20Execute%20a%20high%20impact%20post%20occupancy%20evaluation%20Final.pdf (accessed 14 August 2020)

UK Green Building Council (2016), *Delivering Energy Performance*, https://www.ukgbc.org/wp-content/uploads/2017/09/UK-GBC-Task-Group-Report-Delivering-Building-Performance.pdf (accessed 14 August 2020)

Usable Buildings Trust for guidance, reports and papers, https://www.usablebuildings.co.uk/ (accessed 14 August 2020)

Certification and methodologies

BUS Methodology, https://busmethodology.org.uk/ (accessed 14 August 2020)

Living Building Challenge, https://living-future.org/lbc/ (accessed 14 August 2020)

Well Certification, https://www.wellcertified.com/ (accessed 14 August 2020)

Benchmarks

CarbonBuzz platform for benchmarks, www.carbonbuzz.org/ (accessed 14 August 2020)

CIBSE, *TM46: Energy Benchmarks*, https://www.cibse.org/Knowledge/knowledge-items/detail?id=a0q20000008I7evAAC (accessed 14 August 2020)

CIBSE, *Guide F: Energy Efficiency in Buildings*, https://www.cibse.org/Knowledge/knowledge-items/detail?id=a0q20000008I7oTAAS (accessed 14 August 2020)

Display Energy Certification (DEC), https://www.gov.uk/government/publications/display-energy-certificates-and-advisory-reports-for-public-buildings (accessed 14 August 2020)

LETI Climate Emergency Design Guide (2020), How New Buildings Can Meet UK Climate Change Targets, www.leti.london/publications (accessed 14 August 2020)

Energy Star, https://www.energystar.gov/ (accessed 14 August 2020)

Nabers, https://www.nabers.gov.au/ (accessed 14 August 2020)

Passivhaus, https://www.passivhaustrust.org.uk/ (accessed 14 August 2020)

RIBA 2030 Climate Challenge, https://www.architecture.com/-/media/files/Climate-action/RIBA-2030-Climate-Challenge.pdf?la=en (accessed 14 August 2020)

References

Foreword

1 A Brief Timeline of Climate/UK Policy and Commitments prepared for Futurebuild 2020.
2 B. Bordass and A. Leaman (2013), A New Professionalism: Remedy or Fantasy?, *Building Research & Information*, 41(1), 1–7, doi:10.1080/09613218.2012.750572.
3 LETI (n.d.), Climate Emergency Design Guide, https://www.leti.london/cedg (accessed 1 September 2020).
4 G. Thunberg, L. Neubauer, A. De Wever and A. Charlier (2020), 'After Two Years of School Strikes, the World is Still in a State of Climate Crisis Denial, *Guardian*, https://www.theguardian.com/commentisfree/2020/aug/19/climate-crisis-leaders-greta-thunberg (accessed 1 September 2020).

Introduction

1 Architecture Foundation, 100 Day Studio, 11 June 2020.
2 https://www.architecture.com/knowledge-and-resources/resources-landing-page/riba-plan-of-work (accessed 24 August 2020).

Chapter 1

1 UK Architects Declare Climate and Biodiversity Emergency' https://www.architectsdeclare.com (accessed 24 August 2020).
2 Global Alliance For Buildings and Construction, *2019 Global Status Report for Buildings and Construction Towards a Zero-Emissions, Efficient and Resilient Buildings and Construction Sector*, https://www.unenvironment.org/resources/publication/2019-global-status-report-buildings-and-construction-sector (accessed 24 August 2020).
3 Committee on Climate Change (2019), *Net Zero: The UK's Contribution to Stopping Global Warming*, https://www.theccc.org.uk/publication/net-zero-the-uks-contribution-to-stopping-global-warming/ (accessed 24 August 2020).
4 G. Paterson, D. Mumovic, P. Das and J. Kimpian (2017), 'Energy Use predictions with Machine Learning During Architectural Concept Design', *Science and Technology for the Built Environment*, doi: 10.1080/23744731.2017.1319176; C. van Dronkelaar, M. Dowson, E. Burman, C. Spataru and D. Mumovic (2016), 'A Review of the Energy Performance Gap and Its Underlying Causes in Non-Domestic Buildings', *Frontiers of Mechanical Engineering*, 1(17), doi: 10.3389/fmech.2015.00017.
5 F. Stevenson (2019), *Housing Fit For Purpose: Performance, Feedback and Learning*, RIBA Publishing; The Passivhaus Trust (2017), *The Performance of Passivhaus in New Construction: Post Occupancy Evaluation of Certified Passivhaus Dwellings in the UK: Early Results*, https://www.passivhaustrust.org.uk/UserFiles/File/Technical%20Papers/The%20performance%20of%20Passivhaus%20in%20new%20construction_July%202017%20V2.pdf (accessed 24 August 2020).
6 Innovate UK (2016), Building Performance Evaluation Programme: Findings from Non-Domestic Projects – Getting the Best from Buildings, https://assets.publishing.service.gov.uk/government/uploads/system/uploads/attachment_data/file/497761/Non-Domestic_Building_performance_full_report_2016.pdf (accessed 24 August 2020).
7 Ibid.; Innovate UK (2016) *Building Performance Evaluation Programme: Findings from Domestic projects: Making Reality Match Design*
8 C. Shrubsole, A. Macmillan, M. Davies and N. May (2014), '100 Unintended Consequences of Policies to Improve the Energy Efficiency of the UK Housing Stock', *Indoor and Built Environment*, doi: 10.1177/1420326X14524586.
9 Innovate UK (2016), *Building Performance Evaluation Programme: Findings from Non-Domestic projects – Getting the Best from Buildings*.
10 US Department of Energy (2015), *Quadrennial Technology Review: An Assessment of Energy Technologies and Research Opportunities*, (https://www.energy.gov/sites/prod/files/2017/03/f34/qtr-2015-chapter5.pdf (accessed 24 August 2020).
11 A. Ibrahim and S. Pelsmakers (2018), 'Low-Energy Housing Retrofit in North England: Overheating Risks and Possible Mitigation Strategies', *Building Services Engineering Research and Technology*, 39(2), 161–172, https://doi.org/10.1177/0143624418754386; A. Mavrogianni, M. Davies, J. Taylor, E. Oikonomou, R. Raslan, P. Biddulph, et al. (2013), 'The Unintended Consequences of Energy Efficient Retrofit on Indoor Air Pollution and Overheating Risk in a Typical Edwardian Midterraced House', FutureBuild International Conference; R.S. McLeod, C.J. Hopfe and A. Kwan (2013), 'An Investigation into Future Performance and Overheating Risks in Passivhaus Dwellings', *Building Environment*, 70, 189–209.
12 M. Röck, M.R.M. Saade, M. Balouktsi, F.N. Rasmussen, H. Birgisdottir, R. Frischknecht, et al. (2020), 'Embodied GHG Emissions of Buildings – the Hidden Challenge for Effective Climate Change Mitigation' *Applied Energy*, 258(114107).
13 AIA (2030), 'Commitment by the Numbers', http://content.aia.org/sites/default/files/2019-10/AIA_2030_BytheNumbers_2019.pdf (accessed 24 August 2020); 'RIBA 2030 Challenge', https://www.architecture.com/-/media/files/Climate-action/RIBA-2030-Climate-Challenge.pdf (accessed 24 August 2020).

14 Architects' Council of Europe (2018), *The Architectural Profession in Europe: A Sector Study*, https://www.buildup.eu/sites/default/files/content/2018_ace_report_en_fn.pdf (accessed 24 August 2020).

15 J. Kimpian, E. Burman, J. Bull (2015), *Innovate UK Building Performance Evaluation Non-Domestic Buildings Final Report for Pool and Tremough Innovation Centres*, ref. no. 450043, Innovate UK.

16 O.T. Masoso, L.J. Grobler (2010), 'The Dark Side of Occupants' Behaviour on Building Energy Use', *Energy and Buildings*, 42(2), 173–177; Innovate UK (2016), Building Performance Evaluation Programme: Findings from Non-Domestic Projects – Getting the Best from Buildings.https://assets.publishing.service.gov.uk/government/uploads/system/uploads/attachment_data/file/497761/Non-Domestic_Building_performance_full_report_2016.pdf

Chapter 2

1 F. Musau and K. Steemers (2008), 'Space Planning and Energy Efficiency in Office Buildings: The Role of Spatial and Temporal Diversity', *Architectural Science Review*, 51, 133–145, doi: 10.3763/asre.2008.5117.

2 Quantifiable indicators and recommended and safe levels can be obtained from the World Health Organization (WHO) or country-specific codes and guidance. World Health Organization (2010), *WHO Guidelines for Indoor Air Quality: Selected Pollutants*, WHO Regional Office for Europe.

3 D.P. Wyon and P. Wargocki (2013), 'How Indoor Environment Affects Performance', *ASHRAE Journal*.

4 R. Bunn (2018), *Towards a Theory of Carrying Capacity, Evidence from Long-Term Longitudinal Case Studies of Occupant Satisfaction in Non-Domestic Buildings*, doctoral thesis (Eng.D), University College London.

5 A. Leaman and B. Bordass (2007), 'Are Users More Tolerant of "Green" Buildings?', *Building Research & Information*, 35, 662–673, doi: 10.1080/09613210701529518.

6 Danish Association of Architectural Firms (2020), *Architecture Creates Value*, https://www.danskeark.dk/sites/default/files/2017-06/Architecture_creates_value.pdf (accessed 25 August 2020).

7 Danish Association of Architectural Firms (2020), Architecture Creates Value, https://www.danskeark.dk/sites/default/ files/2017-06/Architecture_creates_value.pdf (accessed 25 August 2020).

8 J. Thomsen, T. Berker, Å.L. Hauge, K. Denizou, S. Wågø and S. Jerkø (2013), 'The Interaction Between Building and Users in Passive and Zero-Energy Housing and Offices: The Role of Interfaces, Knowledge and User Commitment', *Smart and Sustainable Built Environment*, 2(1), 43–59.

9 F. Musau and K. Steemers (2008), 'Space Planning and Energy Efficiency in Office Buildings: The Role of Spatial and Temporal Diversity', *Architectural Science Review*, 51, 133–145, doi: 10.3763/asre.2008.5117.

10 J. Kimpian, E. Burman and S. Chisholm (2013), *Innovate UK Building Performance Evaluation Non-Domestic Buildings Final Report for Petchey, Stockport and Pennywell Academies, Ref no: 1798–16365*, Innovate UK.

11 J. Kimpian, E. Burman and S. Chisholm (2013), Innovate UK Building Performance Evaluation Non-Domestic Buildings Final Report for Petchey, Stockport and Pennywell Academies, Ref no: 1798–16365, Innovate UK.

12 The International WELL Building Institute (2019), *Thermal Comfort*, https://resources.wellcertified.com/articles/what-s-new-in-well-v2-thermal-comfort/ (accessed 25 August 2020).

13 M.A.R. Lopes, C.H. Antunes and N. Martins (2012), 'Energy Behaviours as Promoters of Energy Efficiency: A 21st Century Review', *Renewable and Sustainable Energy Reviews*, 16, 4095–4104, doi: 10.1016/j.rser.2012.03.034; A.R. Hansen (2016), 'The Social Structure of Heat Consumption in Denmark: New Interpretations from Quantitative Analysis', *Energy Research & Social Science*, 11, 109–118, doi: 10.1016/j.erss.2015.09.002; R. Galvin (2013), 'Impediments to Energy-Efficient Ventilation of German Dwellings: A Case Study in Aachen', *Energy and Buildings*, 56, 32–40, doi: 10.1016/j.enbuild.2012.10.020; M. Frontczak, S. Schiavon, J. Goins, et al. (2012), 'Quantitative Relationships Between Occupant Satisfaction and Satisfaction Aspects of Indoor Environmental Quality and Building Design', *Indoor Air*, 22, 119–131, doi: 10.1111/j.1600-0668.2011.00745.x; J. Williams (2017), *A Socio-Technical Method to Assess the Holistic Impact of New Buildings on English Secondary Schools from the Perspective of the Students*, University College London.

14 Lopes, Antunes and Martins (2012), 'Energy Behaviours as Promoters of Energy Efficiency'; M.H. Wu, T.S. Ng and M.R. Skitmore (2016), 'Sustainable Building Envelope Design by Considering Energy Cost and Occupant Satisfaction', *Energy for Sustainable Development*, 31, 118–129, doi: 10.1016/j.esd.2015.12.003.

15 S. Pelsmakers (2015), *The Environmental Design Pocketbook*, RIBA Publishing.

16 European Commission (2016), *An EU Strategy on Heating and Cooling*; T. Parkinson, R.J. de Dear and C. Candido (2012), 'Perception of Transient Thermal Environments: Pleasure and Alliesthesia', *Proceedings of 7th Windsor Conference: The Changing Context of Comfort in an Unpredictable World*, Network for Comfort and Energy Use in Buildings.

17 R.J. de Dear and G. Brager (2001), 'The Adaptive Model of Thermal Comfort and Energy Conservation in the Built

Environment', *International Journal of Biometeorology*, 45, 100–108.
18 A. Wagner, E. Gossauer, C. Moosmann, et al. (2007), 'Thermal Comfort and Workplace Occupant Satisfaction – Results of Field Studies in German Low Energy Office Buildings', *Energy and Buildings*, 39, 758–769, doi: 10.1016/j.enbuild.2007.02.013.
19 Parkinson, de Dear and Candido (2012), 'Perception of Transient Thermal Environments'; Wagner, Gossauer, Moosmann, et al. (2007), 'Thermal Comfort and Workplace Occupant Satisfaction'; de Dear and Brager (2001), 'The Adaptive Model of Thermal Comfort and Energy Conservation in the Built Environment'; R. de Dear, G. Brager and D. Cooper (1997), 'Developing an Adaptive Model of Thermal Comfort and Preference – Final Report', *ASHRAE Journal*, 884.
20 M. Takasu, R. Ooka, H.B. Rijal, et al. (2017), 'Study on Adaptive Thermal Comfort in Japanese Offices Under Various Operation Modes', *Building and Environment*, 118, 273–288, doi: 10.1016/j.buildenv.2017.02.023; S.A. Damiati, S.A. Zaki, H.B. Rijal, et al. (2016), 'Field Study on Adaptive Thermal Comfort in Office Buildings in Malaysia, Indonesia, Singapore, and Japan During Hot and Humid Season', *Building and Environment*, 109, 208–223, doi: 10.1016/j.buildenv.2016.09.024.
21 UK Green Building Council (2016), *Health and Wellbeing in Homes*, https://www.ukgbc.org/wp-content/uploads/2017/12/Healthy-Homes-Full-Report.pdf (accessed 25 August 2020).
22 The International WELL Building Institute (2019), *Thermal Comfort*.
23 S.C. Sekhar (2016), 'Thermal Comfort in Air-Conditioned Buildings in Hot and Humid Climates – Why Are We Not Getting it Right?', *Indoor Air*, 26, 138–152, doi: 10.1111/ina.12184.
24 R.T. Lin and C.C. Chan (2009), 'Effects of Heat on Workers' Health and Productivity in Taiwan', *Global Health Action*, 2, doi:10.3402/gha.v2i0.2024.
25 Sekhar (2016), 'Thermal Comfort in Air-Conditioned Buildings in Hot and Humid Climates'.
26 R.J. de Dear (1998), 'A Global Database of Thermal Comfort Field Experiments', *ASHRAE Journal*, 1141–1152.
27 Parkinson, de Dear and Candido (2012), 'Perception of Transient Thermal Environments'; G. Vargas, R. Lawrence and F. Stevenson (2017), 'The Role of Lobbies: Short-Term Thermal Transitions', *Building Research & Information*, 45, 759–782, doi: 10.1080/09613218.2017.1304095; A.K. Mishra, M.G.L.C. Loomans and J.L.M. Hensen (2016), 'Thermal Comfort of Heterogeneous and Dynamic Indoor Conditions: An Overview', *Building and Environment*, 109, 82–100, doi: 10.1016/j.buildenv.2016.09.016.
28 Vargas, Lawrence and Stevenson (2017), 'The Role of Lobbies: Short-Term Thermal Transitions'.
29 UK Green Building Council (2016), *Health and Wellbeing in Homes*.
30 Parkinson, de Dear and Candido (2012), 'Perception of Transient Thermal Environments'; Vargas, Lawrence and Stevenson (2017), 'The Role of Lobbies: Short-Term Thermal Transitions'; Mishra, Loomans and Hensen (2016), 'Thermal Comfort of Heterogeneous and Dynamic Indoor Conditions'.
31 Vargas, Lawrence and Stevenson (2017), 'The Role of Lobbies: Short-Term Thermal Transitions'.
32 E. Ng (2020), 'From SARS to COVID-19 and Beyond: Public Health Lessons for Buildings', *Buildings and Cities*, https://www.buildingsandcities.org/insights/commentaries/from-sars-to-covid-19-and-beyond.html (accessed 25 August 2020); M. Gormley (2020), 'Pandemics: Why Buildings are Hazards', *Buildings and Cities*, https://www.buildingsandcities.org/insights/commentaries/pandemics-why-buildings-are-hazards.html (accessed 25 August 2020).
33 L. Halonen, E. Tetri and P. Bushel (2010), 'Energy Efficient Electric Lightning for Buildings', ECBCS and Aalto University, http://www.ecbcs.org/Data/publications/EBC_Annex_45_Factsheet.pdf (accessed 25 August 2020).
34 H. Juslén and A. Tenner (2005), 'Mechanisms Involved in Enhancing Human Performance by Changing the Lighting in the Industrial Workplace', *International Journal of Industrial Ergonomics*, 35(9), 843–855, doi:10.1016/j.ergon.2005.03.002.
35 A. Borisuit, F. Linhart, J.-L. Scartezzini and M. Münch (2013), 'Effects of Realistic Office Daylighting and Electric Lighting Conditions on Visual Comfort, Alertness and Mood', *Lighting Research*, 47, 192–209.
36 G.M. Huebner, D.T. Shipworth, S. Gauthier, et al. (2016), 'Saving Energy with Light? Experimental Studies Assessing the Impact of Colour Temperature on Thermal Comfort', *Energy Research & Social Science*, 15, 45–57, doi: 10.1016/j.erss.2016.02.008.
37 International Living Future Institute (2016), *Living Building Challenge*, http://living-future.org/lbc/about (accessed 25 August 2020). The LBC Imperative 10 requires IAQ results one to six months after occupancy: https://living-future.org/wp-content/uploads/2019/08/LBC-4_0_v13.pdf.
38 J. Kimpian, E. Burman and S. Chisholm (2013), *Building Performance Evaluation: Non-Domestic Buildings FinalReport for Loxford School and Brine Leas Sixth Form*; ref. no.1281–16183, Innovate UK. .
39 C. Derix and P. Jagannath (2014), 'Digital intuition – Autonomous Classifiers for Spatial Analysis and Empirical

Design', *The Journal of Space Syntax*, 5(2), 189–215; R. Koenig and G. Schmitt (2016), 'Backcasting and a New Way of Command in Computational Design', https://www.researchgate.net/publication/306525594_Backcasting_and_a_New_Way_of_Command_in_Computational_Design (accessed 25 August 2020).

40 J. Kim and R. de Dear (2013), 'Workspace Satisfaction: The Privacy–Communication Trade-Off in Open-Plan Offices', *Journal of Environmental Psychology*, 36, 18–26, doi: 10.1016/j.jenvp.2013.06.007; K.A. Jensen (2005), 'Acoustical Quality in Office Workstations, as Assessed by Occupant Surveys', *Proceedings: Indoor Air*.

41 D.J. Oldham, M.H. de Salis and S. Sharples (2004), 'Reducing the Ingress of Urban Noise Through Natural Ventilation Openings', *Indoor Air*, 14(8), 118–126, doi:10.1111/j.1600-0668.2004.00294.x; M. Barclay (2012), *The Interaction of Building Energy Use, Ventilation Performance and Urban Noise Under Future Climate Scenarios*, PhD thesis, School of Architecture, The University of Sheffield; M. Barclay, J. Kang, S. Sharples (2012), 'Combining Noise Mapping and Ventilation Performance for Non-Domestic Buildings in an Urban Area', *Building and Environment*, 52, 68–76.

42 J. Kimpian, E. Burman and S. Chisholm (2013), *Building Performance Evaluation: Non-Domestic Buildings Final Report for Loxford School and Brine Leas Sixth Form*; ref. no. 1281–16183, Innovate UK.

43 B. Bordass (2001), *Flying Blind: Everything You Wanted to Know About Energy in Commercial Buildings but Were Afraid to Ask*, EEASOX and the Association for Conservation of Energy (available at: http://www.ukace.org); B. Bordass and A. Leaman (2005), 'Making Feedback and Post-Occupancy Evaluation Routine 1: A Portfolio of Feedback Techniques', *Building Research & Information*, 33(4), 347–352, doi:10.1080/09613210500162016; https://busmethodology.org.uk/ (accessed 25 August 2020).

44 CIBSE (2013), *TM54: Evaluating Operational Energy Performance of Buildings at the Design Stage*, https://www.cibse.org/Knowledge/knowledge-items/detail?id=a0q20000008I7f7AAC (accessed 25 August 2020).

45 London Energy Transformation Initiative, *Climate Emergency Design Guide*, https://www.leti.london/cedg (accessed 25 August 2020).

46 RIBA 2030 Climate Challenge, https://www.architecture.com/-/media/files/Climate-action/RIBA-2030-Climate-Challenge.pdf?la=en (accessed 25 August 2020).

47 J. Kimpian, E. Burman and S. Chisholm (2013), *Building Performance Evaluation: Non-Domestic Buildings FinalReport for Loxford School and Brine Leas Sixth Form*; ref. no.1281–16183, Innovate UK.

Chapter 3

1 For example UKGBC (2020), *Net Zero Carbon: Energy Performance Targets for Offices*, https://www.ukgbc.org/wp-content/uploads/2020/01/UKGBC-Net-Zero-Carbon-Energy-Performance-Targets-for-Offices.pdf (accessed 26 August 2020).

2 For example LETI, *Climate Emergency Design Guide*, https://www.leti.london/cedg (accessed 26 August 2020).

3 T. Lindberg, T. Kaasalainen, M. Moisio, A. Mäkinen, 'M. Hedman and J. Vinha (2020), 'Potential of Space Zoning for Energy Efficiency Through Utilization Efficiency', *Advances in Building Energy Research*, 14(1), 19–40, doi:https://doi.org/10.1080/17512549.2018.1488619.

4 The Usable Buildings Trust, 'Feedback and Strategy for Better-Performing Buildings', https://www.usablebuildings.co.uk/ (accessed 26 August 2020); A. Huovila, P. Tuominen and M. Airaksinen (2017), 'Effects of Building Occupancy on Indicators of Energy Efficiency', *Energies*, 10(5), 628, doi:https://doi.org/10.3390/en10050628.

5 The Usable Buildings Trust, 'Feedback and Strategy for Better-Performing Buildings'.

6 Better Building Partnership (2019), 'Design for Performance', https://www.betterbuildingspartnership.co.uk/node/360 (accessed 26 August 2020); R. Cohen, P. Bannister and B. Bordass (2016), 'Design for Performance Not Compliance: Making Measured Energy In-Use the Objective for New Office Buildings', *International Conference on Energy Efficiency in Commercial Buildings (IEECB '16)*, Frankfurt.

7 CIBSE (2006), *TM22: Energy Assessment and Reporting Methodology*, https://www.cibse.org/Knowledge/knowledge-items/detail?id=a0q20000008I7eWAAS (accessed 26 August 2020).

8 A. Syed (2012), *Advanced Building Technology for Sustainability*, Wiley.

9 Energy Star, 'Energy Star Score', https://portfoliomanager.energystar.gov/pdf/reference/ENERGY%20STAR%20Score.pdf (accessed 26 August 2020).

10 B. Bordass (2020), 'Metrics for Energy Performance in Operation: The Fallacy of Single Indicators', *Buildings and Cities*, 1(1), 260–276, doi: http://doi.org/10.5334/bc.35.

11 CIBSE (2003), *Energy Consumption Guide 19 (ECON19)*, http://pinpoint.ukgbc.org/resource/7897-energy-consumption-guide-19-econ19.php (accessed 26 August 2020).

12 CarbonBuzz, *Using CarbonBuzz*, https://www.carbonbuzz.org/downloads/UsingCarbonBuzz.pdf (accessed 26 August 2020).

13 CIBSE (2012), *Guide F: Energy Efficiency in Buildings*,

https://www.cibse.org/Knowledge/knowledge-items/detail?id=a0q20000008I7oTAAS (accessed 26 August 2020).

14 Better Building Partnership (2019), 'Design for Performance'.

15 Better Buildings Partnership (2016), *Feasibility Study Final Report*, https://www.betterbuildingspartnership.co.uk/sites/default/files/media/attachment/UK%20CAP%20Feasibility%20Study.pdf (accessed 26 August 2020).

16 CIBSE (2012), *Guide F: Energy Efficiency in Buildings*.

17 LETI, *Climate Emergency Design Guide*.

18 RIBA 2030 Climate Challenge, https://www.architecture.com/-/media/files/Climate-action/RIBA-2030-Climate-Challenge.pdf?la=en (accessed 26 August 2020).

19 UKGBC (2020), *Net Zero Carbon: Energy Performance Targets for Offices*.

20 CarbonBuzz, www.carbonbuzz.org (accessed 26 August 2020).

21 Passive House Institute (2016), *Criteria for the Passive House, EnerPHit and PHI Low Energy Building Standard*, https://passiv.de/downloads/03_building_criteria_en.pdf (accessed 26 August 2020).

22 Better Building Partnership (2019), 'Design for Performance'.

Chapter 4

1 D. Jaunzens, P. Grigg, R. Cohen, M. Watson and E Picton (2003), *Building Performance Feedback: Getting Started*, BRE Electronic Publications, https://www.brebookshop.com/details.jsp?id=144574 (accessed 26 August 2020).

2 RIBA (2019), *Sustainable Outcomes Guide*, https://www.architecture.com/-/media/GatherContent/Test-resources-page/Additional-Documents/RIBASustainableOutcomesGuide2019pdf.pdf (accessed 26 August 2020); AIA (2019), *Leveraging Energy Transparency*, http://content.aia.org/sites/default/files/2019-09/EnergyBenchmarking-Whitepaper.pdf (accessed 26 August 2020); K. Settlemyre (2017), 'AIA 2030 Commitment Design Data Exchange', Building Technologies Office Peer Review.

3 B. Bordass, R. Bunn, R. Cohen, M. Standeven and A. Leaman (1997), 'Probe. Lessons Learned from the First Eight Buildings', *CIBSE National Conference*.

4 E. Burman, S. Stamp, N. Jain, C. Shrubsole, D. Mumovic, M. Davies (2018), 'Total Energy and Environmental Performance of Low Carbon Buildings: A Cross Sectoral Study', *CIBSE Technical Symposium*, London.

5 No litigation occurred on the projects included in the BPE programme at Innovate UK 2016. *Building Performance Evaluation Programme: Getting the Best From Buildings*, https://assets.publishing.service.gov.uk/government/uploads/system/uploads/attachment_data/file/497761/Non-Domestic_Building_performance_full_report_2016.pdf (accessed 26 August 2020).

6 J. Kimpian, E. Burman and J. Bull (2015), *Building Performance Evaluation Non-Domestic Buildings Final Report for Pool and Tremough Innovation Centres*, Innovate UK.

7 (RIBA, *Guide to R&D Tax Credits*, https://www.architecture.com/-/media/GatherContent/Test-resources-page/Additional-Documents/12108-RIBARDTax-Creditsv6pdf.pdf accessed 26 August 2020).

8 J. Kimpian, E. Burman, J. Bull, G. Paterson and D. Mumovic (2013), 'Getting Real about Energy Use in Non-Domestic Buildings', accepted for *CIBSE ASHRAE 2014 conference*, London.

9 Ibid.

10 Danish Association of Architectural Firms (2020), *Architecture Creates Value*, https://www.danskeark.dk/sites/default/files/2017-06/Architecture_creates_value.pdf (accessed 26 August 2020).

11 For example, University College London, Oxford Brookes University, Reading University, Plymouth University.

12 R. Hay, S. Bradbury, D. Dixon, K. Martindale, F. Samuel and A. Tait (2016), *Pathways to POE, Value of Architects*, University of Reading, RIBA.

13 Ibid.

14 C. Thoua and J. Hines (2016), 'Post-Occupancy Evaluation of Five Schools by Architype', *Architects Journal*, https://www.architectsjournal.co.uk/buildings/post-occupancy-evaluation-of-five-schools-by-architype/10004526.article?search=https%3a%2f%2fwww.architectsjournal.

Chapter 5

1 A. Stefanoiu (2017), *Towards Performance Evaluation of Energy Efficient Buildings. Construction Durable*, Université Grenoble Alpes; J.S. Haberl, D.E. Claridge and C. Culp (2005), *ASHRAE's Guideline 14-2002 for Measurement of Energy and Demand Savings: How to Determine What Was Really Saved by the Retrofit*, Energy Systems Laboratory, Texas; Passivhaus Trust (2019), Passivhaus: The Route to Zero Carbon? https://www.passivhaustrust.org.uk/UserFiles/File/2019.03.20-Passivhaus%20and%20Zero%20Carbon-Publication%20Version1.2(1).pdf (accessed 27 August 2020); S. Darby (2006), *The Effectiveness of Feedback on Energy Consumption*, Environmental Change Institute, https://www.eci.ox.ac.uk/research/energy/downloads/smart-metering-report.pdf (accessed 27 August 2020); C. Foulds (2013), *Practices and Technological Change: The Unintended Consequences of Low Energy Dwelling Design*, University of East Anglia, School of Environmental Sciences.

2 Building Research Establishment (2018), *Home Quality Mark One, Technical Manual SD239, England, Scotland & Wales*, https://www.homequalitymark.com/wp-content/

uploads/2018/09/HQM-ONE-Technical-Manual-SD239-.pdf (accessed 27 August 2020).

3 S. Pelsmakers, J. Poutanen, S. Saarimaa (2020), '(Hybrid) Architecture In and Over Time', in *Ecologies Design: Transforming Architecture, Landscape and Urbanism*, M.P. Zari, P. Connolly and M. Southcombe (eds), Routledge.

4 Image adapted from J. Futcher and M. Gerald (2014), 'The Selfish Giants', *CIBSE Journal*, http://portfolio.cpl.co.uk/CIBSE/201402/opinion-julie-futcher-gerald-mills/ (accessed 27 August 2020) (2014).

5 'Exploring Masdar City,' http://www.dartmouth.edu/~cushman/courses/engs44-old/Masdar_City.pdf (accessed 27 August 2020).

6 J. Futcher and M. Gerald (2014), 'The Selfish Giants', *CIBSE Journal*, http://portfolio.cpl.co.uk/CIBSE/201402/opinion-julie-futcher-gerald-mills/ (accessed 27 August 2020) (2014).

7 S.-M. Hong, G. Paterson, D. Mumovic, et al. (2013), 'Improved Benchmarking Comparability for Energy Consumption in Schools', *Building Research & Information*, 42, 47–61, doi: 10.1080/09613218.2013.814746.

8 K.J. Lomas and S.M. Porritt (2016), 'Overheating in Buildings: Lessons from Research', *Building Research & Information*, 45, 1–18, doi: 10.1080/09613218.2017.1256136; UKGBC (2016), *Health and Wellbeing in Homes*.

9 B. Smith (2015), *UNISDR Scientific and Technical Advisory Group Case Studies – 2015 – Resilience in UK Schools – Reducing Overheating to Improve Health and Wellbeing*, UNISDR; AECOM (n.d.), *ARISE Case Studies in Disaster Risk Management – Adapting English Schools Sites to Reduce Predicted Overheating Impacts Resulting from Climate Change*.

10 A. Mavrogianni, P. Wilkinson, M. Davies, P. Biddulph and E. Oikonomou (2012), 'Building Characteristics as Determinants of Propensity to High Indoor Summer Temperatures in London Dwellings', *Building and Environment*, 55, 117–130.

11 Zero Carbon Hub (2015), *Overheating in Homes: The Big Picture*.

12 R.S. McLeod and M. Swainson (2017), 'Chronic Overheating in Low Carbon Urban Developments in a Temperate Climate', *Renewable and Sustainable Energy Reviews*, 74, 201–220, doi: 10.1016/j.rser.2016.09.106; C. Morgan, J.A. Foster, A. Poston, et al. (2016), 'Overheating in Scotland: Contributing Factors in Occupied Homes', *Building Research & Information*, 45, 143–156, doi: 10.1080/09613218.2017.1241472; K.J. Beizaee AL and S.K. Firth (2013), National Survey of Summertime Temperatures and Overheating Risk in English Homes, *Building and Environment*, 65, 1–17, doi:10.1016/j.buildenv.2013.03.011.

13 Building Research Establishment (n.d.), *Passivhaus Primer: Designer's Guide – A Guide for the Design Team and Local Authorities*; C. Hopfe, R.S. McLeod (2015), *The Passivhaus Designer's Manual: A Technical Guide to Low and Zero Energy*, Routledge.

14 C. Robertson, 'Productivity, Investment and Energy in a Naturally Ventilated "Lean" Office Development', *CIBSE Technical Symposium, London*.

15 S. Daryanani (1984), 'Design Consideration for the Daylighting of New Commercial Buildings', *Energy and Buildings*, 6, 109–118.

16 Lomas and Porritt (2016), 'Overheating in Buildings: Lessons from Research'; UKGBC (2016), *Health and Wellbeing in Homes*; G. Petrou, P. Symonds, A. Mavrogianni, A. Mylona and M. Davies (2019), 'The Summer Indoor Temperatures of the English Housing Stock: Exploring the Influence of Dwelling and Household Characteristics', *Building Services Engineering Research and Technology*, 40(4), 492–511, doi:https://doi.org/10.1177/0143624419847621.

17 A. Mavrogianni, A. Pathan, E. Oikonomou, P. Biddulph, P. Symonds and M. Davies (2017), 'Inhabitant Actions and Summer Overheating Risk in London Dwellings', *Building Research & Information*, 45(1–2), 119–142, doi:10.1080/09613218.2016.1208431.

18 C. Robertson (2014), *The Role of Crowd-Sourced Energy Performance Feedback in Low-Energy Building Design and Management in the Context of Industry Pressures*, University College London.

19 K. Lomas, R. Giridharan, C.A. Short, A.J. Fair (2012), 'Resilience of "Nightingale" Hospital Wards in a Changing Climate', *Building Services Engineering Research and Technology Journal*, 33, 81–103; E. Gosden (2014), 'Hospitals at increasing risk of overheating due to climate change', *The Telegraph*, http://www.telegraph.co.uk/news/earth/environment/climatechange/10955143/Hospitals-at-increasing-risk-of-overheating-due-to-climate-change.html (accessed 27 August 2020); C. Kendrick, X. Wang, N. Walliman, R. Ogden (2011), 'Evaluation of Ventilation Strategies to Reduce Overheating in a Typical Metal Clad Building With In-Plane Roof-Lights', *International Journal of Architectural Science*, 8(2), 37–46; M. Palme, et al. (2017), 'Natural Ventilation: A Mitigation Strategy to Reduce Overheating in Buildings under Urban Heat Island Effect in South American Cities', *IOP Conference Series: Materials Science and Engineering*.

20 Lomas and Porritt (2016), 'Overheating in Buildings: Lessons from Research'; UKGBC (2016), *Health and Wellbeing in Homes*.

21 D. Ramon, K. Allacker, N.P.M. van Lipzig, F. De Troyer and H. Wouters (2019), 'Future Weather Data for Dynamic

Building Energy Simulations: Overview of Available Data and Presentation of Newly Derived Data for Belgium', in E. Motoasca, A. Agarwal and H. Breesch (eds), *Energy Sustainability in Built and Urban Environments. Energy, Environment, and Sustainability*, Springer; CIBSE, 'Weather Data Sets Packages', https://www.cibse.org/getmedia/628f918c-0fdc-4ddd-9105-4ee7ced9073d/Weath-Data-Sets-Packages-only_1.pdf.aspx (accessed 27 August 2020); S. Belcher, J. Hacker and D. Powell (2005), 'Constructing Design Weather Data for Future Climates', *Building Services Engineering Research and Technology Journal*, 26(1), 49–61, doi:https://doi.org/10.1191/0143624405bt112oa.

22 Image of Clayfields section, adapted from an image by the architects Mikhail Riches, http://www.mikhailriches.com/project/clay-fields/ (accessed 27 August 2020).

23 S. Pelsmakers (2015), *The Environmental Design Pocketbook*, RIBA.

24 S.M. Hong, G. Paterson, D. Mumovic and P. Steadman (2014), 'Improved Benchmarking Comparability for Energy Consumption in Schools', *Building Research & Information*, 42(1), 47–61, doi:10.1080/09613218.2013.814746.

25 G. Vargas, R. Lawrence and F. Stevenson (2017), 'The Role of Lobbies: Short-Term Thermal Transitions', *Building Research & Information*, 45, 759–782, doi:10.1080/09613218.2017.1304095.

26 J. Kimpian, E. Burman and S. Chisholm (2013), *Building Performance Evaluation: Non-Domestic Buildings: Final Report for Petchey, Stockport and Pennywell Academies, ref. no. 1798–16365*, Innovate UK.

27 Ibid.

28 J. Kimpian, E. Burman and S. Chisholm (2013), *Building Performance Evaluation: Non-Domestic Buildings: Final Report for Loxford School and Brine Leas Sixth Form, ref. no. 1281–16183*, Innovate UK.

29 Ibid.

30 Kimpian, Burman and Chisholm (2013), *Building Performance Evaluation: Non-Domestic Buildings: Final Report for Petchey, Stockport and Pennywell Academies*

31 J. Peponis, C. Zimrig and Y.K. Chou (1990), 'Finding the Building in Wayfinding', *Environment and Behaviour*, 22, 550–590.

32 Ibid.

33 Better Buildings Partnership (n.d.), *Green Lease Toolkit*, https://www.betterbuildingspartnership.co.uk/sites/default/files/media/attachment/bbp-gltk-2013_0.pdf (accessed 27 August 2020)

Chapter 6

1 D. van Dijk and J. Hogeling (2019), 'The New ISO 52000 Family of Standards to Assess the Energy Performance of Buildings Put into Practice', *REHVA Journal*, 56(3).

2 S.M. Tabatabaei Sameni, M. Gaterell, A. Montazami, et al. (2015), 'Overheating Investigation in UK Social Housing Flats Built to the Passivhaus Standard', *Building and Environment*, 92, 222–235, doi:10.1016/j.buildenv.2015.03.030; K.J. Beizaee AL and S.K. Firth (2013), 'National Survey of Summertime Temperatures and Overheating Risk in English Homes', *Building and Environment*, 65, 1–17, doi:10.1016/j.buildenv.2013.03.011; A. Mylona and M. Davies (2015), 'Overheating and Indoor Air Quality', *Building Services Engineering Research and Technology Journal* (special issue), 36, 111–114, doi:10.1177/0143624414566551; A. Mavrogianni, P. Wilkinson, M. Davies, et al. (2012), 'Building Characteristics as Determinants of Propensity to High Indoor Summer Temperatures in London Dwellings', *Building and Environment*, 55, 117–130, doi:10.1016/j.buildenv.2011.12.003.

3 F.G.N. Li, A.Z.P. Smith, P. Biddulph, et al. (2014), 'Solid-Wall U-Values: Heat Flux Measurements Compared with Standard Assumptions', *Building Research & Information*, 43, 1–15, doi:10.1080/09613218.2014.967977.

4 J. Wingfield (2010), *Technical Note: Measurement of Heat Loss via Thermal Bypasses in Separating Party Wall Cavities*, Leeds Beckett University; S. Doran (2001), *DETR Framework Project Report: Field Investigations of the Thermal Performance of Construction Elements As Built*, Building Research Establishment; M. Bell, J. Wingfield, D. Miles-Shenton, et al. (2010), *Low Carbon Housing: Lessons from Elm Tree Mews*, Joseph Rowntree Foundation.

5 J. Palmer, D. Godoy-Shimizu, A. Tillson and I. Mawditt (2016), *Building Performance Evaluation Programme: Findings from Domestic Projects: Making Reality Match Design*, Innovate UK.

6 J. Palmer, N. Terry and P. Armitage (2016), *Building Performance Evaluation Programme: Findings from Non-Domestic Projects: Getting the Best from Buildings*, Innovate UK.

7 Palmer, Godoy-Shimizu, Tillson and Mawditt (2016), *Building Performance Evaluation Programme: Findings from Domestic Projects: Making Reality Match Design*.

8 Palmer, Godoy-Shimizu, Tillson and Mawditt (2016), *Building Performance Evaluation Programme: Findings from Domestic Projects: Making Reality Match Design*.

9 Palmer, Godoy-Shimizu, Tillson and Mawditt (2016), *Building Performance Evaluation Programme: Findings from Domestic Projects: Making Reality Match Design*.

10 Building Research Establishment (n.d.), *Passivhaus Primer: Airtightness Guide: Airtightness and Air Pressure Testing in Accordance with the Passivhaus Standard: A Guide for the Design Team and Contractors*, https://www.

passivhaustrust.org.uk/UserFiles/File/Technical%20Papers/BRE_Passivhaus_Airtightness_Guide.pdf (accessed 27 August 2020).

11 bere:architects, https://www.bere.co.uk/research/ (accessed 27 August 2020).

12 O.F. Osanyintola and C.J. Simonson (2006), 'Moisture Buffering Capacity of Hygroscopic Building Materials: Experimental Facilities and Energy Impact', *Energy and Buildings*, 38(10), 1270–1282.

13 N. May (2005), *Breathability: The Key to Building Performance*, http://strawbalebuildinguk.com/wp-content/uploads/2017/09/breathability.pdf (accessed 27 August 2020); N. May and C. Rye (2012), *Responsible Retrofit of Traditional Buildings. A Report on Existing Research and Guidance with Recommendations by STBA*, Sustainable Traditional Buildings Alliance, 1–133, http://www.sdfoundation.org.uk/downloads/RESPONSIBLE-RETROFIT_FINAL_20_SEPT_2012.pdf (accessed 27 August 2020); S. Rhee-Duverne and P. Baker (2013), *Research into the Thermal Performance of Traditional Brick Walls*, English Heritage.

14 May and Rye (2012), *Responsible Retrofit of Traditional Buildings*.

15 S.R. Hastings (2009), *Sustainable Solar Housing: Volume 2: Exemplary Buildings and Technologies*, Routledge.

16 S. Pelsmakers (2015), *The Environmental Design Pocketbook*, RIBA.

17 May (2005), *Breathability: The Key to Building Performance*; May and Rye (2012), *Responsible Retrofit of Traditional Buildings*.

18 A. Bateson (2016), 'Comparison of CIBSE Thermal Comfort Assessments with SAP Overheating Assessments and Implications for Designers', *Building Services Engineering Research and Technology Journal*, 37, 243–251, doi:10.1177/0143624416631133.

19 K. Mithraratne and B. Vale (2004), 'Thermal Characteristics of High Thermal Mass Passive Solar Buildings', *1st International Conference on Sustainability, Engineering and Science*.

20 J. Palmer, D. Godoy-Shimizu, A. Tillson and I. Mawditt (2016), *Building Performance Evaluation Programme: Findings from Domestic Projects: Making Reality Match Design*, Innovate UK.

21 Ibid.

22 Ibid.

23 J. Palmer, N. Terry and P. Armitage (2016), *Building Performance Evaluation Programme: Findings from Non-Domestic Projects: Getting the Best from Buildings*, Innovate UK.

24 C.C. Sullivan (n.d.), *Emerging Trends in Windows, Doors, and Storefronts*, AIA course, https://www.bdcnetwork.com/emerging-trends-windows-doors-and-storefronts-aia-course (accessed 27 August 2020); H. Levin (2003), Designing For People: What Do Building Occupants Really Want?, Healthy Buildings.

25 J. Kimpian, E. Burman and S. Chisholm (2013), *Building Performance Evaluation: Non-Domestic Buildings: Final Report for Petchey, Stockport and Pennywell Academies*, ref. no. 1798–16365, Innovate UK; J. Kimpian, E. Burman and S. Chisholm (2013), *Building Performance Evaluation: Non-Domestic Buildings: Final Report for Loxford School and Brine Leas Sixth Form*, ref. no. 1281–16183, Innovate UK; Palmer, Godoy-Shimizu, Tillson and Mawditt (2016), *Building Performance Evaluation Programme: Findings from Domestic Projects: Making Reality Match Design*; Palmer, Terry and Armitage (2016), *Building Performance Evaluation Programme: Findings from Non-Domestic Projects: Getting the Best from Buildings*.

26 Kimpian, Burman and Chisholm (2013), *Building Performance Evaluation: Non-Domestic Buildings: Final Report for Petchey, Stockport and Pennywell Academies*; Palmer, Terry and Armitage (2016), *Building Performance Evaluation Programme: Findings from Non-Domestic Projects: Getting the Best from Buildings*.

27 J. Kimpian, E. Burman and J. Bull, *Building Performance Evaluation: Non-Domestic Buildings: Final Report for Pool and Tremough Innovation Centres*, ref. no. 450043, Innovate UK; A. Raj, D. Hes, R. Padovani and C. Jensen (2010), 'Building Satisfaction – Using Thermal Modelling to Identify Areas of 2 Building Use Focus For Post Occupancy Evaluation', *Proceedings of Building Simulation 2011: 12th Conference of International Building Performance Simulation Association*.

28 J. Kimpian, D. Rigamonti, G. Paterson (2012), *Design for a Future Climate: Harris Academy, Purley*, https://www.arcc-network.org.uk/wp-content/D4FC/D4FC17-Harris-academy-full-report.pdf (accessed 27 August 2020).

29 A. López, F.D. Molina-Aiz, D.L. Valera and A. Peña (2016), 'Wind Tunnel Analysis of the Airflow through Insect-Proof Screens and Comparison of Their Effect When Installed in a Mediterranean Greenhouse', *Sensors*, 16(5), 690, doi:10.3390/s16050690.

30 Kimpian, Burman and Chisholm (2013), *Building Performance Evaluation: Non-Domestic Buildings: Final Report for Petchey, Stockport and Pennywell Academies*; Kimpian, Burman and Chisholm (2013), *Building Performance Evaluation: Non-Domestic Buildings: Final Report for Loxford School and Brine Leas Sixth Form*; Palmer, Godoy-Shimizu, Tillson and Mawditt (2016), *Building Performance Evaluation Programme: Findings from Domestic Projects: Making Reality Match Design*; Palmer, Terry and Armitage (2016), *Building Performance Evaluation Programme: Findings from Non-Domestic Projects: Getting the Best from Buildings*.

31 Kimpian, Burman and Chisholm (2013), *Building Performance Evaluation: Non-Domestic Buildings: Final Report for Loxford School and Brine Leas Sixth Form*; Palmer, Terry and Armitage (2016), *Building Performance Evaluation Programme: Findings from Non-Domestic Projects: Getting the Best from Buildings*.

Chapter 7

1 Berkeley Lab (2017,) Indoor Air Quality Scientific Findings Resource Bank – Increased Use of Air Conditioning, https://iaqscience.lbl.gov/cc-ac (accessed 28 August 2020).
2 S.-M. Hong, G. Paterson, D. Mumovic, et al. (2013), 'Improved Benchmarking Comparability for Energy Consumption in Schools', *Building Research & Information*, 42: 47–61, doi:10.1080/09613218.2013.814746.
3 J. Kimpian and S. Watts (2010), 'A Blueprint for the Future Office', *Council on Tall Buildings and Urban Habitat Conference*.
4 Adapted from E. Burman (2016), 'Building Performance Gap: A Way Forward', *Building Performance Evaluation – Understanding the Performance Gap*, University College London.
5 J. Kimpian and S. Watts (2010), 'A Blueprint for the Future Office', *Council on Tall Buildings and Urban Habitat Conference*.
6 J. Hummelgaard, P. Juhl, K.O. Sæbjörnsson, et al. (2007), 'Indoor Air Quality and Occupant Satisfaction in Five Mechanically and Four Naturally Ventilated Open-Plan Office Buildings', *Building and Environment*, 42, 4051–4058, doi:10.1016/j.buildenv.2006.07.042.
7 S. Finlay, I. Pereira, E. Fryer-Smith, A. Charlton and R. Roberts-Hughes (2012), *The Way We Live Now: What People Need and Expect from Their Homes, A Research Report for the Royal Institute of British Architects*, RIBA, https://www.ipsos.com/sites/default/files/publication/1970-01/sri-riba-the-way-we-live-now-may-2012.pdf (accessed 28 August 2020).
8 S.M. Joshi (2008), 'The Sick Building Syndrome', *Indian Journal of Occupational Environmental Medicine*, 12(2), 61–64, doi:10.4103/0019-5278.43262; United States Environmental Protection Agency, 'An Office Building Occupants Guide to Indoor Air Quality', https://www.epa.gov/indoor-air-quality-iaq/office-building-occupants-guide-indoor-air-quality#design (accessed 28 August 2020).
9 J. Kimpian, E. Burman and S. Chisholm (2013), *Building Performance Evaluation: Non-Domestic Buildings: Final Report for Petchey, Stockport and Pennywell Academies*, ref. no. 1798–16365, Innovate UK.
10 Ibid.
11 E. Ng (2020), 'From SARS to COVID-19 and Beyond: Public Health Lessons for Buildings', *Buildings and Cities*, *https://www.buildingsandcities.org/insights/commentaries/from-sars-to-covid-19-and-beyond.html* (accessed 28 August 2020).
12 European Centre for Disease Prevention and Control (2020), 'Heating, Ventilation and Air-Conditioning Systems in the Context of COVID-19', https://www.ecdc.europa.eu/sites/default/files/documents/Ventilation-in-the-context-of-COVID-19.pdf (accessed 28 August 2020).
13 Kimpian, Burman and Chisholm (2013), *Innovate UK Building Performance Evaluation Non-domestic Buildings Final Report for Petchey, Stockport and Pennywell Academies*.
14 E. Burman, S. Stamp, N. Jain, C. Shrubsole, D. Mumovic and M. Davies (2018), 'Total Energy and Environmental Performance of Low Carbon Buildings: A Cross Sectoral Study', *CIBSE Technical Symposium*, *https://discovery.ucl.ac.uk/id/eprint/10047043/1/CIBSE_Symposium_2018_Paper_184_FINAL_180221.pdf* (accessed 28 August 2020); CIBSE (n.d.), *TM61: An Integrated Framework for Building Performance*.
15 J. Kimpian, E. Burman and S. Chisholm (2013), *Building Performance Evaluation: Non-Domestic Buildings: Final Report for Loxford School and Brine Leas Sixth Form*, ref. no. 1281–16183, Innovate UK; J. Palmer, N. Terry and P. Armitage (2016), *Building Performance Evaluation Programme: Findings from Non-Domestic projects – Getting the Best from Buildings*, Innovate UK.
16 Kimpian, Burman and Chisholm (2013), *Building Performance Evaluation: Non-Domestic Buildings: Final Report for Loxford School and Brine Leas Sixth Form.*
17 J. Kimpian, E. Burman and J. Bull (2015), *Building Performance Evaluation: Non-Domestic Buildings: Final Report for Pool and Tremough Innovation Centres*, ref. no. 450043, Innovate UK.
18 K. de Selincourt (2015), 'Mechanical ventilation and IAQ – what the evidence reveals', *Passive House+*, *https://passivehouseplus.ie/magazine/insight/mechanical-ventilation-and-iaq-what-the-evidence-reveals?highlight=WyJrYXRIIl0=* (accessed 28 August 2020).
19 E. Burrell (2015), 'What is Mechanical Ventilation with Heat Recovery?', *Passivhaus in Plain English and More* (blog), *https://elrondburrell.com/blog/passivhaus-mechanical-ventilation-heat-recovery/* (accessed 28 August 2020).
20 P. Dunbabin and C. Wickins (2012), *Detailed Analysis from the First Phase of the Energy Saving Trust's Heat Pump Field Trial*, Energy Saving Trust UK, *https://assets.publishing.service.gov.uk/government/uploads/system/uploads/attachment_data/file/48327/5045-heat-pump-field-trials.pdf* (accessed 28 August 2020); P. Dunbabin, H. Charlick and R. Green (2013), *Detailed Analysis from the Second Phase of the Energy Saving Trust's Heat Pump Field Trial*, Energy Saving Trust UK, https://assets.

publishing.service.gov.uk/government/uploads/system/uploads/attachment_data/file/225825/analysis_data_second_phase_est_heat_pump_field_trials.pdf (accessed 28 August 2020).

21 Dunbabin, Charlick and Green (2013), *Detailed Analysis from the Second Phase of the Energy Saving Trust's Heat Pump Field Trial*; Dunbabin, Charlick and Green (2013), *Detailed Analysis from the Second Phase of the Energy Saving Trust's Heat Pump Field Trial*.

22 Kimpian, Burman and Chisholm (2013), *Innovate UK Building Performance Evaluation Non-domestic Buildings Final Report for Petchey, Stockport and Pennywell Academies*.

23 Palmer, Terry and Armitage (2016), *Building Performance Evaluation Programme: Findings from Non-Domestic projects – Getting the Best from Buildings*; N. Nahar (2002), Capital Cost and Economic Viability of Thermosyphonic Solar Water Heaters Manufactured from Alternate Materials in India', *Renew Energy*, 26, 623–635; A. Khalifa (1998), 'Forced Versus Natural Circulation Solar Water Heaters: A Comparative Performance Study', *Renew Energy*, 14, 77–82.

24 S. Pelsmakers (2015), *The Environmental Design Pocketbook*, RIBA.

25 A. Rentizelas, A. Tolis and I.P. Tatsiopoulos (2009), Logistics Issues of Biomass: The Storage Problem and the Multi-Biomass Supply Chain, *Renewable & Sustainable Energy Reviews*, 13(4), 887–894; U.A. Schneider and B.A. McCarl (2001), 'Economic Potential of Biomass-Based Fuels for Greenhouse Gas Emission Mitigation', *CARD Working Papers*, 307, *http://lib.dr.iastate.edu/card_workingpapers/307* (accessed 28 August 2020); D.J. Immerzeel, P.A. Verweij, F. van der Hilst, A.P.C. Faaij (2014), 'Biodiversity Impacts of Bioenergy Crop Production: A State-of-the-Art Review', *GCB Bioenergy*, 6, 183–209, doi:http://dx.doi.org/10.1111/gcbb.12067; A. Gasparatos, C.N.H. Doll, M. Esteban, A. Ahmed and T.A. Olang (2017), 'Renewable Energy and Biodiversity: Implications for Transitioning to a Green Economy', *Renewable Sustainable Energy Review*, 70, 161–184, doi:10.1016/j.rser.2016.08.030.

26 Kimpian, Burman and Bull (2015), *Building Performance Evaluation: Non-Domestic Buildings: Final Report for Pool and Tremough Innovation Centres*.

27 L. Halonen, E. Tetri and P. Bushal (2010), *Guidebook on Energy Efficient Electric Lightning for Buildings*, ECBCS and Aalto University, http://www.ecbcs.org/Data/publications/EBC_Annex_45_Factsheet.pdf (accessed 28 August 2020).

28 CIBSE (2001), *Lighting Guide 11: Surface Reflectance and Colour*, https://www.cibse.org/knowledge/knowledge-items/detail?id=a0q20000008I7kLAAS (accessed 28 August 2020).

29 Palmer, Terry and Armitage (2016), *Building Performance Evaluation Programme: Findings from Non-Domestic projects – Getting the Best from Buildings*.

30 A. Menezes, A. Cripps, D. Bouchlaghem and R. Buswell (2012), 'Predicted vs. Actual Energy Performance of Non-Domestic Buildings, Using Post-Occupancy Evaluation Data to Reduce the Performance Gap', *Applied Energy*, 97, 355–364; C. Van Dronkelaar, M. Dowson, C. Spataru, D. Mumovic (2016), A Review of the Energy Performance Gap and Its Underlying Causes in Non-Domestic Buildings', *Frontiers in Mechanical Engineering*,1, 1–14, doi:http://doi.org/10.3389/fmech.2015.00017; IPEEC (2019), *International Review of Building Energy Performance Gap Issues*, http://cbd.gov.au/files/IPEEC%20BEET%20Performance%20Gap%20Report%20Final%20Draft%208Oct2019.pdf (accessed 28 August 2020); BREEAM (2018), *GN32: BREEAM New Construction Guidance Note – Energy Prediction and Post Occupancy Assessment*, https://www.breeam.com/wp-content/uploads/sites/3/2018/04/GN32_BREEAM_UKNC_2018_Energy_prediction_and_post_occupancy_assessment_v0.0.pdf (accessed 28 August 2020).

31 Kimpian, Burman and Chisholm (2013), *Innovate UK Building Performance Evaluation Non-domestic Buildings Final Report for Petchey, Stockport and Pennywell Academies*.

32 J. Kimpian, D. Rigamonti and G. Paterson (2012), *Harris Academy, Purley*, https://www.arcc-network.org.uk/wp-content/D4FC/D4FC17-Harris-academy-full-report.pdf (accessed 28 August 2020).

33 Ibid.

Chapter 8

1 E. Burman (2016), 'Building Performance Gap: A Way Forward', *Building Performance Evaluation – Understanding the Performance Gap*, University College London.

2 J. Palmer, D. Godoy-Shimizu, A. Tillson and I. Mawditt (2016), *Building Performance Evaluation Programme: Findings From Domestic Projects: Making Reality Match Design*, Innovate UK.

3 E. Burman (2016), 'Building Performance Gap: A Way Forward', *Building Performance Evaluation – Understanding the Performance Gap*, University College London.

4 J. Kimpian, E. Burman and S. Chisholm (2013), *Building Performance Evaluation: Non-domestic Buildings: Final Report for Loxford School and Brine Leas Sixth Form*, ref. no. 1281–16183, Innovate UK.

5 J. Kimpian, E. Burman and J. Bull (2015), *Building Performance Evaluation: Non-Domestic Buildings: Final Report for Pool and Tremough Innovation Centres*, ref. no. 450043, Innovate UK.

6 G. Gao and K. Whitehouse (2009), 'The Self-Programming Thermostat: Optimizing Setback Schedules Based on Home Occupancy Patterns', *Proceedings of the First ACM Workshop on Embedded Sensing Systems for Energy-Efficiency in Buildings*, 67–72, https://10.1145/1810279.1810294 (accessed 28 August 2020); E. Soltanaghaei and K. Whitehouse (2017), 'Practical Occupancy Detection for Programmable and Smart Thermostats', *Applied Energy*, doi:http://dx.doi.org/10.1016/j.apenergy.2017.11.024.
7 M. Hannus, A.S. Kazi, Z. Alain (n.d.), *ICT Supported Energy Efficiency in Construction: Strategic Research Roadmap and Implementation Recommendations*, https://ec.europa.eu/information_society/activities/sustainable_growth/docs/sb_publications/reeb_ee_construction.pdf (accessed 28 August 2020).
8 M. Brambley, D. Hansen, P. Haves, D. Holmberg, S. McDonald, K. Roth and P. Torcellini (2005), 'Advanced Sensors and Controls for Building Applications: Market Assessment and Potential R&D Pathways', technical report, Pacific Northwest National Laboratory.
9 Ibid.
10 Kimpian, Burman and Chisholm (2013), *Building Performance Evaluation: Non-domestic Buildings: Final Report for Loxford School and Brine Leas Sixth Form.*
11 J. Kimpian, E. Burman and S. Chisholm (2013), *Building Performance Evaluation: Non-Domestic Buildings: Final Report for Petchey, Stockport and Pennywell Academies*, ref. no. 1798–16365, Innovate UK.
12 Kimpian, Burman and Chisholm (2013), *Building Performance Evaluation: Non-domestic Buildings: Final Report for Loxford School and Brine Leas Sixth Form.*
13 Ibid.
14 Feedback undertaken at Oxford Brookes University: https://www.brookes.ac.uk/architecture/research/low-carbon-building-group/building-performance-evaluation/college-lake/ (accessed 28 August 2020).
15 Kimpian, Burman and Bull (2015), *Building Performance Evaluation: Non-Domestic Buildings: Final Report for Pool and Tremough Innovation Centres.*
16 Kimpian, Burman and Chisholm (2013), *Building Performance Evaluation: Non-domestic Buildings: Final Report for Loxford School and Brine Leas Sixth Form.*
17 Kimpian, Burman and Chisholm (2013), *Building Performance Evaluation: Non-Domestic Buildings: Final Report for Petchey, Stockport and Pennywell Academies.*
18 BCO, *201 Bishopsgate: Post Occupancy Evaluation – An Occupier's Perspective*, http://www.bco.org.uk/Research/Publications/Post_Occupancy_Evaluation_-_An_Occupiers_Perspective.aspx (accessed 28 August 2020).
19 Kimpian, Burman and Bull (2015), *Building Performance Evaluation: Non-Domestic Buildings: Final Report for Pool and Tremough Innovation Centres.*
20 Ibid.
21 Kimpian, Burman and Chisholm (2013), *Building Performance Evaluation: Non-domestic Buildings: Final Report for Loxford School and Brine Leas Sixth Form.*
22 Kimpian, Burman and Bull (2015), *Building Performance Evaluation: Non-Domestic Buildings: Final Report for Pool and Tremough Innovation Centres.*
23 Directive 2012/27/EU of the European Parliament and of the Council of 25 October 2012 on energy efficiency, amending Directives 2009/125/EC and 2010/30/EU and repealing Directives 2004/8/EC and 2006/32/EC Text with EEA relevance, https://eur-lex.europa.eu/legal-content/EN/TXT/?qid=1399375464230&uri=CELEX:32012L0027 (accessed 28 August 2020).
24 VITO and Waide (2020), *3rd Interim Report of the 2nd Technical Support Study on the Smart Readiness Indicator for Buildings*, https://smartreadinessindicator.eu/sites/smartreadinessindicator.eu/files/sri2-_third_interim_report.pdf (accessed 28 August 2020).
25 Ibid.

Chapter 9

1 D. Hawkes (2008), *The Environmental Imagination*, Routledge.
2 T. Schulze and E. Eicker (2013), 'Controlled Natural Ventilation for Energy Efficient Buildings', *Energy and Buildings*, https://doi.org/10.1016/j.enbuild.2012.07.044 (accessed 29 August 2020).
3 Compiled and analysed by Carnegie Mellon's Center for Building Performance and Diagnostics.
4 Live data feeds in real time can be viewed on the RMI website: https://rmi.org/our-work/buildings/scaling-zero-net-carbon/rmi-innovation-center/ (accessed 29 August 2020).
5 Conducted by Lawrence Berkeley Labs.
6 Undertaken by Aarhus University researchers.
7 Henning Larsen's sustainability department working with the Technical University of Denmark, University of Copenhagen, Aarhus University and the Danish Building Research Institute.
8 Danish Building Regulations, https://bygningsreglementet.dk/Historisk/BR18_Version3/Tekniske-bestemmelser/11/BRV/Funktionsafprovning/Indledning (accessed 30 August 2020).
9 CIBSE (2012), *Guide F: Energy Efficiency in Buildings.*
10 RIBA Plan of Work 2020, https://www.architecture.com/-/media/GatherContent/Test-resources-page/Additional-Documents/2020RIBAPlanofWorktemplatepdf.pdf (accessed 31 August 2020).

Chapter 10

1 Example of an energy risk register: https://www.maxfordham.com/assets/media/images/publications/

Keynsham%20Civic%20Centre/Keynsham%20 Publications%20Download%202017.pdf (accessed 31 August 2020).

2 From correspondence with Diana Dina, Haworth Tompkins, head of sustainability and regenerative design, June 2020.

3 https://www.architecture.com/knowledge-and-resources/resources-landing-page/post-occupancy-evaluation (accessed 31 August 2020).

4 F. Stevenson (2019), *Housing Fit for Purpose: Performance, Feedback and Learning*, RIBA.

5 RIBA Plan of Work 2020, https://www.architecture.com/-/media/GatherContent/Test-resources-page/Additional-Documents/2020RIBAPlanofWorktemplatepdf.pdf (accessed 31 August 2020).

6 C. Shrubsole, et al. (2019), Bridging the Gap: The Need for a Systems Thinking Approach in Understanding and Addressing Energy and Environmental Performance in Buildings, *Indoor and Built Environment*, 28(1), 100–117.

7 RIBA 2030 Climate Challenge https://www.architecture.com/-/media/files/Climate-action/RIBA-2030-Climate-Challenge.pdf?la=en (accessed 31 August 2020).

8 European Commission (n.d.), *Level(s): The European Framework for Sustainable Buildings*, https://ec.europa.eu/environment/eussd/buildings.htm (accessed 31 August 2020).

9 CIBSE (2013), *TM54: Evaluating Operational Energy Performance of Buildings at the Design Stage*, https://www.cibse.org/Knowledge/knowledge-items/detail?id=a0q20000008I7f7AAC (accessed 31 August 2020).

10 https://www.ashrae.org/technical-resources/bookstore/standard-90-1 (accessed 31 August 2020).

11 https://www.techstreet.com/ashrae/standards/ashrae-90-2-2018?product_id=2030773 (accessed 31 August 2020).

12 BPIE, (n.d.), *Building Renovation Passports: Consumer's Journey to a Better Home*, http://bpie.eu/publication/building-renovation-passports-consumers-journey-to-a-better-home/ (accessed 31 August 2020).

13 For different BIM levels, see https://www.thenbs.com/knowledge/bim-levels-explained (accessed 31 August 2020).

Glossary

1 Air Tightness Testing & Measurement Association (ATTMA) standards TSL1 and TSL2, https://www.bcta.group/product/test-standards/ (accessed 14 August 2020); Passivhaus Trust UK, *Passivhaus Primer: Airtightness Guide*, https://www.passivhaustrust.org.uk/UserFiles/File/Technical%20Papers/BRE_Passivhaus_Airtightness_Guide.pdf (accessed 14 August 2020).

2 N-G. Vardaxis, D. Bard, K. Persson Waye (2017), 'On the definition of acoustic comfort in residential buildings', *Acoustical Society of America Journal*, 141(5), p 3540, https://doi.org/10.1121/1.4987481.

3 B. Bordass, R. Cohen, P. Bannister (2016), *UK Commitment Agreements: Making Measured Energy In-Use the Objective for New Office Buildings*, Feasibility Study Final Report, https://www.betterbuildingspartnership.co.uk/sites/default/files/media/attachment/UK%20CAP%20Feasibility%20Study%20Final%20Report%2017May16.pdf (access 14 August 2020).

4 M. MacDonald, S. Livengood (2000), *Benchmarking Residential Energy Use*, http://amet-me.mnsu.edu/userfilesshared/solarwall/Benchmarking/Benchmarking%20Residential%20Energy%20Use.pdf (accessed 14 August 2020); W. Chung, Y.V. Hui, Y.M. Lam (2006), 'Benchmarking the Energy Efficiency of Commercial Buildings', *Applied Energy*, 83(1), pp 1–14.,https://www.sciencedirect.com/science/article/abs/pii/S0306261904002028 (accessed 14 August 2020).

5 Britannica.com, 'Biophilia hypothesis', https://www.britannica.com/science/biophilia-hypothesis (access 14 August 2020); E. O. Wilson (1984), *Biophilia*, Harvard University Press.

6 A.K.H. Kwan and P.L. Ng (2015), 'Building Diagnostic Techniques and Building Diagnosis: The Way Forward', in P. Tse, J. Mathew, K. Wong, R. Lam, C. Ko (eds), *Engineering Asset Management: Systems, Professional Practices and Certification*, Lecture Notes in Mechanical Engineering, Springer.

7 E. Giama and A.M. Papadopoulos (2012), 'Sustainable Building Management: Overview of Certification Schemes and Standards', *Advances in Building Energy Research*, 6(2), pp 242–258, doi: 10.1080/17512549.2012.740905.

8 J. Sinopoli (2010), 'Chapter 12: Facility Management Systems', in *Smart Building Systems for Architects, Owners and Builders*, J. Sinopoli (ed.), Butterworth-Heinemann, https://doi.org/10.1016/B978-1-85617-653-8.00012-0.

9 Innovate UK, *Building Performance Evaluation Programme: Findings from Domestic Projects Making Reality Match Design*, https://assets.publishing.service.gov.uk/government/uploads/system/uploads/attachment_data/file/497758/Domestic_Building_Performance_full_report_2016.pdf (access 14 August 2020); Innovate UK, *Building Performance Evaluation Programme: Findings from Non-Domestic Projects*, https://assets.publishing.service.gov.uk/government/uploads/system/uploads/attachment_data/file/497761/Non-Domestic_Building_performance_full_report_2016.pdf (accessed 14 August 2020).

10 W. Bordass (2001) *Flying Blind: Everything You Wanted To Know About Energy in Commercial Buildings But Were Afraid To Ask*, EEASOX and the Association for Conservation of Energy, available at http://www.ukace.

org; B. Bordass and A. Leaman (2005), 'Making Feedback and Post-Occupancy Evaluation Routine 1: A Portfolio of Feedback Techniques', *Building Research & Information*, 33(4), pp 347–352, doi: 10.1080/09613210500162016; https://busmethodology.org.uk/ (accessed 14 August 2020).
11 CarbonBuzz, https://www.carbonbuzz.org/docs/CarbonBuzz_Handbook.pdf (accessed 14 August 2020).
12 https://www.cibse.org/ (accessed 14 August 2020).
13 D. Farhud, Z. Aryan (2018), 'Circadian Rhythm, Lifestyle and Health: A Narrative Review', *Iran Journal of Public Health*, 47(8), pp 1068–1076, https://www.ncbi.nlm.nih.gov/pmc/articles/PMC6123576/ (accessed 14 August 2020).
14 EA (2019), *Tracking the Decoupling of Electricity Demand and Associated CO_2 Emissions*, IEA, Paris https://www.iea.org/commentaries/tracking-the-decoupling-of-electricity-demand-and-associated-co2-emissions (accessed 14 August 2020).
15 S. Pelsmakers (2015) *The Environmental Design Pocketbook*, RIBA Publishing.
16 G.J. Levermore (2000), *Building Energy Management Systems: Applications to Low Energy HVAC and Natural Ventilation Control*, second edition, E&FN Spon.
17 C.W. Gellings (1985), 'The Concept of Demand-Side Management for Electric Utilities', in *Proceedings of the IEEE*, 73(10), pp 1468–1470, doi: 10.1109/PROC.1985.13318.
18 Department for Communities and Local Government (2015), *Improving the Energy Efficiency of Our Buildings*, https://assets.publishing.service.gov.uk/government/uploads/system/uploads/attachment_data/file/452481/DEC_Guidance__rev_July_2015_.pdf (accessed 14 August 2020).
19 M. De Rosa, V. Bianco, F. Scarpa, L.A. Tagliafico (2016), 'Modelling of Energy Consumption in Buildings: An Assessment of Static and Dynamic Models', *Russian Journal of Construction Science and Technology*, 2(1), pp 12–24; A. Foucquier, S. Robert, F. Suard, L. Stephan, A. Jay (2013), 'State of the Art In-Building Modelling and Energy Performances Prediction: A Review', *Renewable and Sustainable Energy Reviews*, 23(272), p 288, doi: 10.1016/j.rser.2013.03.004.
20 UK Green Building Council (2017) , 'Embodied Carbon: Developing a Client Brief', https://www.ukgbc.org/sites/default/files/UK-GBC%20EC%20Developing%20Client%20Brief.pdf (accessed 14 August 2020).
21 Domestic Energy Performance Certificate Register, https://www.epcregister.com/ (accessed 14 August 2020); A. Arcipowska, F. Anagnostopolous, F. Mariottini, S. Kunkel, *Energy Performance Certificates Across the EU: A Mapping of National Approaches*, BPIE, http://bpie.eu/wp-content/uploads/2015/10/Energy-Performance-Certificates-EPC-across-the-EU.-A-mapping-of-national-approaches-2014.pdf (accessed 14 August 2020).
22 Energy Performance of Buildings Directive, https://ec.europa.eu/energy/topics/energy-efficiency/energy-efficient-buildings/energy-performance-buildings-directive_en (accessed 14 August 2020).
23 X. Deng, G. Kokogiannakis, Z. Ma and P. Cooper (2017), 'Thermal Comfort Evaluation of a Mixed-Mode Ventilated Office Building with Advanced Natural Ventilation and Underfloor Air Distribution Systems' (2017), Faculty of Engineering and Information Sciences – Papers: Part A, p 6642, https://ro.uow.edu.au/eispapers/6642 (accessed 14 August 2020); A. Leaman and B. Bordass (2007), 'Are Users More Tolerant of "Green" Buildings?', *Building Research & Information*, 35(6), pp 662–673.
24 A. Esteves, M. Esteves, M. Mercado, G. Barea and D. Gelardi (2018), 'Building Shape that Promotes Sustainable Architecture, Evaluation of the Indicative Factors and Its Relation with the Construction Costs', *Architecture Research*, 8(4), pp 111–122, http://dx.doi.org/10.5923/j.arch.20180804.01; R. McLeod, K. Mead and M. Standen, 'Passivhaus Primer – Designer's Guide: A Guide for the Design Team and Local Authorities', BRE Trust, https://passivehouse-international.org/upload/BRE_Passivhaus_Designers_Guide.pdf (accessed 14 August 2020).
25 Pelsmakers (2015), *The Environmental Design Pocketbook*.
26 A. Leaman, F. Stevenson and B. Bordass (2010), 'Building Evaluation: Practice and Principles', *Building Research & Information*, 38:5, pp 564–577, http://dx.doi.org/10.1080/09613218.2010.495217.
27 Pelsmakers (2015), *The Environmental Design Pocketbook*.
28 United States Environmental Protection Agency (EPA), 'Indoor Air Quality', https://www.epa.gov/indoor-air-quality-iaq/introduction-indoor-air-quality (accessed 14 August 2020).
29 Pelsmakers (2015), *The Environmental Design Pocketbook*.
30 A.P.C. Chan and A.P.L. Chan (2004), 'Key Performance Indicators for Measuring Construction Success', *Benchmarking: An International Journal*, 11(2), pp 203–221, https://doi.org/10.1108/14635770410532624.
31 European Commission, *Level(s): A Guide to Europe's New Reporting Framework for Sustainable Buildings*, https://ec.europa.eu/environment/eussd/pdf/Level_publication_EN.pdf (accessed 14 August 2020).
32 Pelsmakers (2015), *The Environmental Design Pocketbook*.

33 International Living Future Institute (2019), *Living Building Challenge 4.0: A Visionary Path to a Regenerative Future*, https://living-future.org/wp-content/uploads/2019/04/Living-Building-Challenge-4.0.pdf (accessed 14 August 2020).
34 G. Brager (2006), 'Mixed-Mode Cooling', *ASHRAE Journal*, August, pp 30–37.
35 https://www.nabers.gov.au/ (accessed 14 August 2020).
36 M. De Simone, C. Carpino, D. Mora, S. Gauthier, V. Aragon, G.U. Harputlugil (2018), 'Reference Procedures for Obtaining Occupancy Profiles in Residential Buildings', https://annex66.org/sites/default/files/2018FinalReport/Subtask%20A%20Deliverable%20-%20Reference%20procedures%20for%20obtaining%20occupancy%20profiles%20in%20residential%20buildings.pdf (accessed 14 August 2020).
37 Ibid.
38 R. Azari (2019), 'Chapter 5: Life Cycle Energy Consumption of Buildings: Embodied + Operational', in V. Tam, L. Khoa (eds), *Sustainable Construction Technologies*, Butterworth-Heinemann, https://doi.org/10.1016/B978-0-12-811749-1.00004-3.
39 O. Fanger (1970), *Thermal Comfort*, Danish Technical Press.
40 R. McMullan (2007), *Environmental Science in Building,* sixth edition, Palgrave Macmillan.
41 CIBSE, PROBE Studies, https://www.cibse.org/knowledge/probe-post-occupancy-studies (accessed 14 August 2020).
42 Pelsmakers (2015), *The Environmental Design Pocketbook*.
43 BSRIA, Soft Landings, https://www.bsria.com/uk/consultancy/project-improvement/soft-landings/ (accessed 14 August 2020).
44 M. De Rosa, et al. (2016), 'Modelling of Energy Consumption in Buildings'; A. Foucquier, et al. (2013), 'State of the Art In-Building Modelling and Energy Performances Prediction'.
45 Pelsmakers (2015), *The Environmental Design Pocketbook*.
46 Ibid.
47 Ibid.
48 Ibid.
49 Ibid.
50 CIBSE, *TM54: Evaluating Operational Energy Performance of Buildings at the Design Stage*, https://www.cibse.org/Knowledge/knowledge-items/detail?id=a0q20000008I7f7AAC (accessed 14 August 2020).
51 Ibid.
52 Pelsmakers (2015), *The Environmental Design Pocketbook*.
53 Ibid.
54 D.B. Lu, D.M. Warsinger (2020), 'Energy Savings of Retrofitting Residential Buildings with Variable Air Volume Systems Across Different Climates', *Journal of Building Engineering*, Elsevier BV, 30(101223), doi:10.1016/j.jobe.2020.101223.
55 Pelsmakers (2015), *The Environmental Design Pocketbook*.
56 Ibid.
57 Ibid.

Index

Note: page numbers in *italics* refer to figures; page numbers in **bold** refer to tables; page numbers followed by an asterisk refer to side texts

A

academic collaboration 45
acoustic comfort 24
  impact on other comfort parameters 25*
  open-plan spaces 24, 59
  outdoor noise 55, 73, 80
  parameters and indicators affecting **17**
acoustic louvres *72*, *73*, *81*, 158, *159*
adaptive comfort 21, 22, 72, 124
additional services 177
adjacency 57, 59
AHMM
  Morelands Rooftop, London *54*, 71*, 122–9
  White Collar Factory, Old Street, London *83*
AHR architects 46*
  Al Bahr Towers, Abu Dhabi *71*
  Keynsham Civic Centre, Keynsham, Somerset *26*, *57*, 71*, *72*, *84*, 89, *99*, 154–61
air recirculation 23, 82
air-change rates 23, 72, 82
air-conditioning 79, 117
air-handling units (AHUs) 12
airtightness 66
  affect on indoor air quality 65, 82
  detailing 66, 68, 136
  regulatory context 65
Al Bahr Towers, Abu Dhabi *71*
architects' involvement 13, 45
Architype
  Enterprise Centre, University of East Anglia, Norwich 85, *94*, 99*, 103*, 162–9
  Wilkinson Primary School, Wolverhampton 46*, 47

B

background ventilation 82, 85
backup boilers 88*, 124
baseline energy performance 40, 156, 180
benchmarks 25, 33, **35**
bere:architects
  Mayville Community Centre, London *67*, *103*
  Passfield Drive, London, UK *67*
Better Buildings Partnership
  Design for Performance 30*, 34
  Green Lease 60–1
biomass 88
blinds 71
  automated 98, 124, 150
  user operation 71*, 132, 135, 160
Bloomberg, City of London *81*
'breathable' building envelope 68
BREEAM 9, 25, 122
buffer zones 58
building configuration 52–9, 91, **175**
building contracts *see* performance contracts; procurement
building energy use diagnostics 30, 44
building envelope *see* fabric performance
building form 52, 55
building management systems (BMS) 98–106
  communication issues 103–5
  energy saving 99
  procurement 105–6
  sensors and set points 101
  smart systems 106
  (sub) metering 102–3
  and user experience (UX) design 106
  zoning and intermittent occupancy 100
building massing 52, 55
building orientation 57–8
building performance evaluation (BPE) 4, 39
  costs 43–4, 104–5
  expertise and skills 45
  and procurement 43
  and professional indemnity (PI) insurance 42
Building Performance Register (BPR) 173–81
building performance targets *see* performance targets
building regulations 40, 54, 69, 90, 98, 150, 153
Building Schools for the Future 52
building services 77–93
building usage 13, 20, 59
Building Use Studies (BUS) 25
  benchmarking parameters 25
  case studies 128–9, 136, 166, 169
building user manuals *103*, 161

C

carbon emissions *see* CO_2 emissions
carbon intensity 9, 29–30, 33*, 87, 88*
carbon neutral *see* zero carbon buildings
CarbonBuzz 33, **35**, 46*
Carrowbreck Meadow, Norfolk, UK *69*, 71*, *86*, 130–7
case studies 111–69
  Carrowbreck Meadow, Norfolk, UK *69*, 71*, *86*, 130–7
  Enterprise Centre, University of East Anglia, Norwich 85, *94*, 99*, 103*, 162–9
  Everyman Theatre, Liverpool *6*, *62*, 85–6, 88*, 114–21, 174*
  Frederiksbjerg School, Aarhus, Denmark *48*, 59, 146–53
  Keynsham Civic Centre, Keynsham, Somerset *26*, *57*, 71*, *72*, *84*, 89, *99*, 154–61

Morelands Rooftop, London *54*, 71*, 122–9
Rocky Mountain Institute, Colorado, USA 98, 138–45, *170*
ceiling heights 54, 55*, 80
certification systems 9
see also compliance calculations/certification models
CIBSE guidance documents 25, 30, 33, **35**, 191
circadian rhythms 21, 23
circulation areas 58
climate change resilience 52, 82*, 176
CO_2 emissions 9, 10, 29
CO_2 monitoring 82, 85*, 101
coefficient of performance (CoP) 87
cognitive performance 18
combined heat and power (CHP) 88
commissioning 92, 104, 105*, 179
communication processes 103–4
compliance calculations/certification models 40–1, 65, 68, 69, 90
see also Energy Performance Certification (EPC)
condensation 67
contractual requirements *see* performance contracts
control systems 95–109
blinds *see under* blinds
sensors and sensor locations 101
set points and system profiles 101
smart controls 106–7
stakeholder involvement *105*
windows *see under* windows
zoning 100
see also building management systems (BMS); lighting controls
cooling 84–5
heat pumps 87
labyrinths and earth tubes 85–6
mechanical systems 55, 59, 79
see also night-purge ventilation/cooling
Cork House, Berkshire, UK *66*
COVID-19 23, 82

D
data gathering and reporting 43–4
see also Building Performance Register (BPR)
data privacy issues 34
daylight dimming *89*, 90
daylighting 23, 71, 80
demandside management (DSM) 107
design and build (D&B) 12, 13*, 43
design changes 12
see also value engineering
design coordination 91
Design for Future Climate Programme 55*
Design for Performance 30*, 34
detailing 66, 68, 135
Display Energy Certification (DEC) 10, 11, 33*, **35**
district heating 88
documentation *103*, 106, 161
domestic hot water 87
doors 70–1, 73, 98*
ductwork openings 70
dynamic energy models 41, 69

E
earth tubes 85–6
ECON19 33
Eddington Nursery, Cambridge *22*
electric vehicle charging 107
electricity consumption 30, 31, 33, 79
see also energy use; grid electricity
electricity transmission and conversion losses 31
embodied carbon 12, 29, 55*
energy bar diagram *31*
energy intensity 29
energy metering 102–3
meter reconciliation 30
submetering 30*, 88, 102
Energy Performance Certification (EPC) 40, 109
Energy Performance of Buildings Directive (EPBD) 107
energy saving 99
energy sources 30
Energy Star 11, **35**
energy use
baseline 40, 156, 180
benchmarks 33, **35**
'by fuel' or 'by end use' 30
daily energy profiles 32
heat kWhs vs electricity kWhs 31
levels of detail and their use *34*
metrics 29
modelling 41–2
operational energy frameworks and standards 34
performance targets **175**
regulatory framework 11–12
reporting 29, 30
unregulated energy use *40*, 41–2, 90
volumetric reporting 29*
vs carbon emissions 29
enhanced commissioning 179
Enterprise Centre, University of East Anglia, Norwich 85, *94*, 99*, 103*, 162–9
equipment and appliance loads *see* unregulated energy use
Everyman Theatre, Liverpool *6*, *62*, 85–6, 88*, 114–21, 174*
external blinds 71
automated 98, 124, 150
user operation 71*, 132
external evaluators 43
external louvres 71

external noise 55, 73, 80
external pollutants 23, 55
  electric vehicles impact 80
  and naturally ventilated buildings 84
external shading 52*, 55, 71, 118, 132, 178
  see also external blinds

F

fabric performance 65–75, **175**
feedback
  barriers to 42–7
  defined 4
  hierarchy *51*
  public disclosure 46
feedback loops 3, 13, 44
feedback process *2–3*, 39, 44–5
  architects' involvement 13, 45
  costs 43–4, 104–5
  expertise and skills 45
flexibility score 19*
floor heights 54, 55*, 80
forgiveness factor 19*, *54*, 80
form factor 52
Foster + Partners
  Bloomberg, City of London *81*
  Masdar, Abu Dhabi *53*
Frederiksbjerg School, Aarhus, Denmark *48*, 59, 146–53
fresh air rates *see* air-change rates
fuel CO_2 intensity factors 9, 29–30, 33*, 87, 88*
fuel consumption 30, 31
  see also energy use
functional proximity 59

G

gas-fired boilers 88*, 124
glare 23, 58, 71
glazing 73
  see also windows
graduated response approach 39, *180*
Green Lease 60–1
grid electricity, carbon intensity 9, 87, 88*

H

Hamson Barron Smith, Carrowbreck Meadow, Norfolk, UK *69*, 71*, *86*, 130–7
Haworth Tompkins, Everyman Theatre, Liverpool *6*, *62*, 85–6, 88*, 114–21, 174*
heat pumps 85, 87, 117–18
heating and hot water systems 87–8, 97
Henning Larsen, Frederiksbjerg School, Aarhus, Denmark *48*, 59, 146–53
holistic approach 12
  see also whole-life carbon emissions
hot water *see* domestic hot water
human factor 13
  see also occupant behaviour; usability; UX (user experience) design
humidity 68
HVAC systems 79, 80, 82
  see also air-conditioning; heating and hot water systems; ventilation
hygroscopic materials 68

I

indoor air quality (IAQ) 17–18
  monitoring 19*, *84*, 85*, 101
  occupant satisfaction 19, *24*, 25*
  parameters and indicators affecting **17**, 24, 25*
  sealed buildings 65, 82
  underperformance 10
Indoor Environmental Quality (IEQ) 17
  affecting cognitive performance and productivity 18
  forgiveness factor 19, *54*, 80
  parameters and indicators affecting **17**, 24, 25*
  performance targets **175**
  see also indoor air quality (IAQ)
innovative systems 12, 91–2
insulation 65, 67
  continuity 68, *69*
  delamination 67
  installation problems 70
  vapour-impermeable 68
inter-disciplinary working 91
internal layout 57–61
internal pollutants 23
interstitial condensation 67

J

JTP offices, Pennington Warehouse, London *14*, *18*, 19*

K

key performance indicators (KPIs) 39, 109
Keynsham Civic Centre, Keynsham, Somerset *26*, *57*, 71*, *72*, *84*, 89, *99*, 154–61

L

labyrinths 85–6
LED lighting 90
LEED 25, 138
lessons learned 46
light transmittance 73
lighting 23, 89–90
  case studies 152–3
  and thermal comfort perception 23*
  visual comfort **17**, 23
  see also daylighting

lighting controls
daylight dimming *89*, 90
occupancy detection 90, 100*, 101
Living Building Challenge (LBC) 22, 23*
London Energy Transformation Initiative (LETI), Climate Emergency Design Guide 25, **35**
low carbon buildings *see* zero carbon buildings
low-carbon energy metering 102*
low-carbon heating 87, 88

M

M&E systems *see* technical systems
manual controls/operation 97
blinds 71, 124, 135, 141
windows 72–3, 98*, 124, 141
Masdar, Abu Dhabi *53*
massing 52, 55
Mayville Community Centre, London *67*, *103*
mechanical cooling 55, 59, 79
mechanical ventilation with heat recovery (MVHR) 85, *86*, 135
mechanically ventilated buildings 79–80, 82
meter reconciliation 30, 102
metering 30*, 88, 102–3
microclimate 52
Mikhail Riches, Clay Field, Suffolk, UK *56*
mixed-mode ventilation 80, 85, 118
moisture movement 68, 127
Morelands Rooftop, London *54*, 71*, 122–9
mosquito gauze 85
MVHR *see* mechanical ventilation with heat recovery (MVHR)

N

NABERS (National Australian Built Environment Rating System) 11, 25, **35**
named subcontractors 12
naturally ventilated buildings 78, 82–5
case studies 114, 117–18, 124
control systems 101
forgiveness factor 19*
occupant satisfaction 80
vs mechanically ventilated buildings 79–80
window design 72
see also night-purge ventilation/cooling
net zero energy *see* zero carbon/energy
night-purge ventilation/cooling 69, 84–5, 127, 141
NO_2 levels *84*
noise 24, 25*, 55, 73, 80
non-fixed equipment 90
novation 12, 43

O

occupancy detection 90, 100*, 101
occupancy patterns 13*, 58–9
and energy intensity reporting 29
and energy use 100
and smart controls 106
occupant behaviour 13, 20
blind operation 71, 135, 160
CO_2 monitoring 101*
compromising natural ventilation 84
window opening 73, 84, 128
occupant controls 21, 55, 97
blinds 71, 124, 135
windows 98*, 124, 141
occupant satisfaction 19
and IEQ measurements *24*
impact of design quality 19
naturally ventilated buildings 80
performance targets **175**
surveying 19–20
see also forgiveness factor
occupant surveys *see* Building Use Studies (BUS)
openings 70–1
see also doors; windows
open-plan spaces 24, 59
operating conditions 13, 41, 90, 91, 180*
operational energy frameworks 34
operational ratings *see* Display Energy Certification (DEC)
orientation of building 57–8
outdoor pollutants *see* external pollutants
out-of-hours use 59–61
overheating risk 10, 52, 54–5, 58, 73, 91, 128
Oxford Brookes University 101*

P

Passfield Drive, London, UK *67*
passive cooling 84–5
see also naturally ventilated buildings; night-purge ventilation/cooling
passive design 51
Passivhaus 11, **35**, 66, *67*, 85, *103*
in case studies 132, 168
performance contracts 10, 156
see also Building Performance Register (BPR)
performance gap 4, 9–11, *31*
by building type *11*
complex buildings *97*
energy modelling improvements 41–2
fabric issues 69–70
naturally vs mechanically ventilated buildings *80*
potential reduction 42
unregulated energy use affecting 90

performance indicators *see* key performance indicators (KPIs)
performance reconciliation 108–9
performance targets 25, 39, 173–4, **175**, 181
  see also benchmarks; performance contracts
performance validation 10, 180*
pipe openings 70
pollutants 23
  see also external pollutants
post-occupancy evaluation (POE) 4
  see also building performance evaluation (BPE)
predicted mean vote (PMV) 21
pre-project surveys 176
primary energy use 31
procurement 12, 43, 105–6
productivity gains 18, 19, 144
professional indemnity (PI) insurance 42
public disclosure 46

Q
quality assurance 107*, 108–9

R
reconciliation of performance 108–9
reflectance, surface 89–90
regulated and unregulated energy 41–2, 90–1
regulatory framework 11–12
  see also building regulations
renewable energy 29, 91
respiratory comfort **17**, 23
RIBA
  2030 Climate Challenge 25, **35**, 128
  Plan of Work *2–3*, *174*
risk registers 173
Rocky Mountain Institute, Colorado, USA 98, 138–45, *170*
rooflights 70–1

S
sealed buildings 79, 82
security grilles 85
sensitivity analyses 177
sensors and sensor locations 101
service openings 70–1
shading *see* external shading
site context 52, 56–7
smart controls 98*, 106–7
smart homes 106*
Smart Readiness Indicators (SRIs) 107
Soft Landings 25, 34, 42, 99*, 105*
  in case studies 156, 166
solar design 52, 56
solar gain 71, 73
solar shading *see* external shading
solar water heaters (SWHs) 88
solid wall construction 65
space heating 87, 97
spatial design 57–61, 91
specialist expertise 45, 92
steady-state models 41, 69
Storey's Field Centre, Cambridge *76*
submetering 30*, 88, 102
surface reflectance 89–90
switches 97
  see also manual controls/operation
systems integration 77–93, **175**

T
task lighting 89
technical systems 77–93
tenanted buildings 30*, 34
terminology 4
thermal alliesthesia 21–2
thermal breaks 68
thermal bridges 68, *69*, 72
thermal comfort 17–18
  architectural qualities affecting 23*
  forgiveness factor 19*, *54*, 80
  occupant perception 19, 25*, 54, 55*
  parameters and indicators affecting **17**, 21–2, 25*
  temperature range 21
  underperformance 10
  variable conditions 21–2
  see also adaptive comfort
thermal mass 68–9
  case studies 117, 119, 127, 141, 143, 157
  existing buildngs 55*, 65
  overheating risk 69, 127
thermal performance 65–75
  performance targets **175**
  regulatory context 65
  troubleshooting 69–70
thermal pleasure 21–2
thermal resistance 65, 67
thermal transmittance (U-value) 65, 73
thermal zoning 57, 58–9, 59–60, 100
thermostats 97, 98*
transitional spaces 58
trees 55*, 132, 178
triple glazing 73

U
UEA Enterprise Centre 73*
UK Committee for Climate Change *10*
unregulated energy use *40*, 41–2, 90–1
urban design 52, 57
urban heat island 52

usability
  building management systems (BMS) 103–5, 106
  manually operated windows 72–3
  switches 97
  window controls 98*
user experience (UX) design 51, 91, 96, 106, **175**
user manuals *103*, 161
user-centric design 13
U-values 65, 73
UX (user experience) design 13

V

value engineering 12, 22*, 91, 178, 179
vapour permeability 68
variable-speed fans 82
ventilation 23, 78, 79, *80*
  control systems 101, *102*
  mechanical systems 79–80, 82
  night-purge ventilation 69, 84–5, 127, 141
  see also air-change rates; naturally ventilated buildings; windows
views 23
visual comfort **17**, 23
  see also glare; lighting
visual transmittance 73

W

weather data 55*
WELL Building Standard 22, 25
wellbeing 17–18
White Collar Factory, Old Street, London *83*
whole-life carbon emissions 12, 29, 88
whole-life costs 80, 93
Wilkinson Primary School, Wolverhampton 46*, 47
windows
  automated opening 13*, 73, *81*
  design affecting air flow 72
  design considerations 70–1
  obstructions to air flow 85
  openable 72–3, 84, 98*, 124, 128, 141
  security considerations 85
  size and position 72
  triple glazing 73
workmanship 66, 68, 135

Z

zero carbon buildings 12, 29, 91, 105*, 107
zero carbon/energy, terminology 29
ZGF with Graybeal Architects, Rocky Mountain Institute, Colorado, USA 98, 138–45, *170*
zoning *see* thermal zoning

Image credits

Page II Architype / Dennis Gilbert **Page IV** AHR / Daniel Hopkinson **Figure 0.1** Henning Larsen / Hufton + Crow **Figure 0.2** *Energy People Buildings* authors **Figure 1.1** Tim Soar **Figure 1.2** *Energy People Buildings* authors **Figure 1.3** Committee for Climate Change Report 2019 – redrawn for *Energy People Buildings* **Figure 1.4** Innovate UK Building Performance Evaluation Programme: Findings from non-domestic projects – redrawn for *Energy People Buildings* **Figure 2.1** JTP / Fotohaus / Craig Auckland **Figure 2.2** JTP / Fotohaus / Craig Auckland **Figure 2.3** *Energy People Buildings* authors redrawn from a trademark of, and used here with permission from, International WELL Building Institute, pbc. **Figure 2.4** Alan Williams **Figure 2.5** AHR – redrawn for *Energy People Buildings* **Figure 3.1** AHR / Daniel Hopkinson **Figure 3.2** *Energy People Buildings* authors **Figure 3.3** *Energy People Buildings* authors **Figure 3.4** Dr Esfand Burman – redrawn for *Energy People Buildings* **Figure 3.5** *Energy People Buildings* authors **Figure 4.1** Hamson Barron Smith / Jefferson Smith **Figure 4.2** *Energy People Buildings* authors **Figure 4.3** *Energy People Buildings* authors **Figure 4.4** Architype / Dennis Gilbert **Figure 5.1** Henning Larsen / Hufton + Crow **Figure 5.2** *Energy People Buildings* authors **Figure 5.3** Julie Futcher – adapted for *Energy People Buildings* **Figure 5.4** Foster + Partners / Nigel Young **Figure 5.5** AHMM – adapted for *Energy People Buildings* **Figure 5.6** *Energy People Buildings* authors **Figure 5.7** Mikhail Riches **Figure 5.8** *Energy People Buildings* authors / Getmapping plc, Infoterra Ltd & Bluesky, Maxar Technologies, map data 2020 **Figure 6.1** Haworth Tompkins **Figure 6.2a** CSK / Matthew Barnett Howland, Dido Milne, Oliver Wilton **Figure 6.2b** CSK / Magnus Dennis **Figure 6.3** Bere Architects **Figure 6.4** Bere Architects **Figure 6.5** Hamson Barron Smith **Figure 6.6** *Energy People Buildings* authors **Figure 6.7** AHR / Christian Richters **Figure 6.8** AHR / Daniel Hopkinson / Dr Esfand Burman (thermal image) **Figure 7.1** MUMA / Jim Stephenson **Figure 7.2** *Energy People Buildings* authors **Figure 7.3** Authors of *Sustainably Tall: Investment Energy Life Cycle* – redrawn for *Energy People Buildings* **Figure 7.4** Dr Esfand Burman – redrawn for *Energy People Buildings* **Figure 7.5** Foster + Partners / Nigel Young **Figure 7.6** Authors of *Sustainably Tall: Investment Energy Life Cycle* – redrawn for Energy People Buildings **Figure 7.7** AHMM / Tim Soar **Figure 7.8** Dr Sam Stamp / UCL IEDE – redrawn for *Energy People Buildings* **Figure 7.9** Hamson Barron Smith **Figure 7.10** Judit Kimpian **Figure 8.1** Dennis Gilbert **Figure 8.2** *Energy People Buildings* authors **Figure 8.3** Judit Kimpian **Figure 8.4** Judit Kimpian **Figure 8.5** AHR – redrawn for *Energy People Buildings* **Figure 8.6** AHR – redrawn for *Energy People Buildings* **Figure 8.7** Bere Architects **Figure 8.8** *Energy People Buildings* authors **Figure 8.9** *Energy People Buildings* authors **Figure 9.1.1** Haworth Tompkins **Figure 9.1.2** Haworth Tompkins **Figure 9.1.3** Haworth Tompkins **Figure 9.1.4** Haworth Tompkins **Figure 9.1.5** Haworth Tompkins **Figure 9.1.6** Haworth Tompkins **Figure 9.1.7** Haworth Tompkins **Figure 9.1.8** *Energy People Buildings* authors **Figure 9.2.1** AHMM **Figure 9.2.2** AHMM / Tim Soar **Figure 9.2.3** AHMM **Figure 9.2.4** AHMM **Figure 9.2.5** AHMM / Tim Soar **Figure 9.2.6** *Energy People Buildings* authors **Figure 9.2.7** AHMM / Rob Parrish **Figure 9.2.8** *Energy People Buildings* authors **Figure 9.3.1** Hamson Barron Smith **Figure 9.3.2** Hamson Barron Smith / Jefferson Smith **Figure 9.3.3** Hamson Barron Smith **Figure 9.3.4** Hamson Barron Smith / Jefferson Smith **Figure 9.3.5** Hamson Barron Smith / Jefferson Smith **Figure 9.3.6** Hamson Barron Smith / *Energy People Buildings* authors **Figure 9.3.7** Hamson Barron Smith / *Energy People Buildings* authors **Figure 9.3.8** *Energy People Buildings* authors **Figure 9.4.1** ZGF Architects **Figure 9.4.2** ZGF Architects / Tim Griffiths **Figure 9.4.3** DHM Design **Figure 9.4.4** ZGF Architects **Figure 9.4.5** ZGF Architects / Tim Griffiths **Figure 9.4.6** ZGF Architects – adapted for *Energy People Buildings* **Figure 9.4.7** RMI – adapted for *Energy People Buildings* **Figure 9.4.8** RMI – adapted for *Energy People Buildings* **Figure 9.5.1** Henning Larsen / Hufton + Crow **Figure 9.5.2** Henning Larsen / Hufton + Crow **Figure 9.5.3** Henning Larsen / Hufton + Crow **Figure 9.5.4** Henning Larsen – adapted for *Energy People Buildings* **Figure 9.5.5** Henning Larsen **Figure 9.5.6** Henning Larsen / Imke Wies Mulder-van Mil **Figure 9.5.7** Henning Larsen / Imke Wies Mulder-van Mil **Figure 9.5.8** *Energy People Buildings* authors **Figure 9.6.1** AHR / Daniel Hopkinson **Figure 9.6.2** AHR **Figure 9.6.3** AHR **Figure 9.6.4** AHR **Figure 9.6.5** AHR **Figure 9.6.6** AHR **Figure 9.6.7** AHR **Figure 9.6.8** *Energy People Buildings* authors **Figure 9.7.1** Architype **Figure 9.7.2** Architype / Dennis Gilbert **Figure 9.7.3** Architype / Darren Carter and Morgan Sindall **Figure 9.7.4** Architype **Figure 9.7.5** Architype / Darren Carter and Morgan Sindall **Figure 9.7.6** *Energy People Buildings* authors **Figure 9.7.7** Architype / Darren Carter and Morgan Sindall **Figure 9.7.8** *Energy People Buildings* authors / Architype **Figure 10.1** DHM Design **Figure 10.2** *Energy People Buildings* authors **Figure 10.3** *Energy People Buildings* authors **Figure 10.4** *Energy People Buildings* authors **Figure 11.1** Haworth Tompkins